The Entrepreneurial
Engineer

The Entrepreneurial Engineer

Personal, Interpersonal, and Organizational Skills for Engineers in a World of Opportunity

David E. Goldberg
The University of Illinois at Urbana–Champaign

A JOHN WILEY & SONS, INC., PUBLICATION

For general information on our other products and services or for technical support, please contact our Customer Care Department within the United States at (800) 762-2974, outside the United States at (317) 572-3993 or fax (317) 572-4002.

Wiley also publishes its books in a variety of electronic formats. Some content that appears in print may not be available in electronic formats. For more information about Wiley products, visit our web site at www.wiley.com.

Library of Congress Cataloging-in-Publication Data is available.

ISBN-13 978-0-470-00723-5
ISBN-10 0-470-00723-0

Printed in the United States of America.

10 9 8 7 6 5 4 3 2

To the penguins

Contents

6 Present, Don't Speak   74

7 Human Side of Engineering   91

Foreword

If there was ever a time for this book it is now. With the United States facing continued pressure to follow the global outsourcing trend, the country must adapt to maintain its edge as a global innovator. But how?

A cultural renaissance that effects the venerable educational and corporate institutions is required. Technology, whether it is bits and bytes or biosciences, has pervaded society and continues to drive progress. The traditional role of the number-crunching or code-writing engineer, who was employed by the thousands by large corporation, is quickly becoming extinct. Trends such as Open Innovation, in which Fortune 1,000 companies are increasingly looking outside for innovation and new intellectual property, are taking hold across the country.

Dr. Goldberg provides the roadmap for engineers of the future to stay at the front of the wave by learning to think more like entrepreneurs. By definition the entrepreneur is accustomed to—if not thrives in—a messy world where innovation and application intersect to produce transformation of products, companies, and industries. Required skills include understanding the business impact of the innovation and communicating ideas in a clear and compelling manner.

However, skills are not enough. The main message of this book—and the secret that Dr. Goldberg is conveying—is that the passion for the idea, doing what you love, and having the persistence required to bring ideas to reality are the fuel of innovation. Without them, the world will not change, and the idea will stay in the notebook.

Consider this book your survival handbook for th erest of your life. Onward and upward!

Tim Schigel
Cincinnati, Ohio

Preface

This book and its predecessor *Life Skills and Leadership for Engineers* (Goldberg, 1995), owe their writing to my first engineering job out of school at what was then a small entrepreneurial engineering-software firm called Stoner Associates, Inc., in Carlisle, Pennsylvania. I remember thinking before I started how technically exciting it was going to be to work on four complex hydraulics computer programs. I remember wondering shortly thereafter if other engineers spent so much of their time writing, talking, selling, meeting, and worrying about building our customer base. At one time, I thought my experience at Stoner Associates was a bit extreme, but I've come to realize that it was a precursor of the experiences that more and more engineers face every year. Today's engineer is challenged by a hectic pace at work, direct involvement in business matters, and by a large amount of personal contact, meetings, reports, and presentations.

The key thesis of this book is that today's engineers are confronted by and expected to participate in a world of opportunity, opportunity driven by technical change, enabled by financial and business prowess, and implemented through skilled, collaborative labors. Although it was never right to send engineers out into the workforce thinking that math, science, and design were the totality of an engineer's portfolio, to do so in today's marketplace is a form of educational malpractice. Fortunately, there is a growing awareness of the need to alert engineers early in their education and in their careers to the importance of mastering a balance of technical and nontechnical skills, and this book aims at the latter kind with the goal of creating a new breed of *entrepreneurial engineers* who mix strong technical skills, business and organizational prowess, and an alert eye to opportunity.

The book is designed for use in a classroom setting or for self-study in a variety of ways: (1) in a one-credit course to be given to undergraduate engineering students a year or so before they graduate, (2) as a supplement to an introductory engineering class for underclass students, (3) as a supplement to a three- or four-credit senior capstone course, (4) as a supplement to a graduate seminar, (5) as part of course materials for corporate training or continuing education for engineers, and (6) for self-study by engineers and their mentors. No particular prerequisites are assumed, although upon occasion common engineering terms are taken for granted, and mathematical ideas from first-year calculus are used.

ACKNOWLEDGMENTS

Many people provided much important feedback during the preparation of the original volume, *Life Skills and Leadership for Engineers* (LLE), and the present volume, *The Entrepreneurial Engineer* (TEE).

XV

For help with LLE, I thank Bob Barfield, Larry Bergman, John Biddle, David Brown, Jim Carnahan, Dermot Collins, Harry Cook, Chuck Crider, Sydney Cromwell, Kalyanmoy Deb, Pete DeLisle, Jerry Dobrovolny, Chuck Evces, Anju Jaggi, Rachel Janssen, Don Jones, Ed Kuznetsov, Angie Locascio, James McDonough, Dan Metz, Dave O'Bryant, Constantine Papadakis, Carolyn Reed, Tom Richwine, Kate Sherwood, Angus Simpson, Mark Spong, Mark Strauss, Christy Tidd, Debbie Thurston, and Ben Wylie.

For help with TEE, I thank Doug Bosworth, Camila Chaves, Jeff Cochran, Kathy Jamison, Xavier Llora, Chris Magee, Joel Moses, Roscoe Pershing, Courtney Price, Ray Price, Kumara Sastry, and Tian-Li Yu.

I thank my wife, Mary Ann, for her love, steadfast support, and for a last-minute reading of the whole manuscript. My sons Max and Zack are growing into delightful young men; they were mere children during the writing and publication of LLE. I thank Max for his humor, his courage, and a skillful proofreading of a number of TEE chapters. I thank Zack for teaching me the meaning and importance of a hug.

Chapter 1

Entrepreneurial Engineer: Ready for the 21st Century

1.1 21ST-CENTURY ENGINEERS MOVING AT INTERNET TIME

The Internet. Global markets. Time compression. Competitiveness. We increasingly live in a connected world where packets of information whiz around the globe at the speed of light, carrying the electronic equivalent of letters, money, contracts, designs, or other work product between actors located anywhere on the planet. It is against such a backdrop that engineers plot a course for a life of rewarding, productive work, but it is a world vastly different from the one faced by engineers even a few short years ago. Not long ago, the engineer of the Cold War prepared for work by immersing himself in a narrow technical discipline, expecting to work his entire career for one of a small number of gigantic employers on some specialized subsystem of a defense-related or smokestack megaproject. Today's engineer is on a different planet. He or she faces a life filled with multiple project assignments with an almost interchangeable array of employers, clients, startups, and established firms; these assignments require an extraordinarily broad set of technical, business, and interpersonal skills performed as part of ever-changing and shifting interdisciplinary teams.

This change in work life has been as rapid as it has been dramatic, and the job here is to survey those nontechnical skills essential to being a successful engineer in the 21st century. As opposed to the Cold War engineer, we call the ideal engineer of our times an *entrepreneurial engineer*, and here we interpret the word *entrepreneur* quite broadly.

In the traditional sense of the word, today's engineer *is* more likely to find him or herself as part of a startup, replete with 13-hour workdays, a Blackberry, and stock options. But even when today's engineer works in more traditional settings, he or she is likely to find that both the job itself and effective career management require a more venturesome attitude and approach. Increased competition places enormous pressure on companies to continue to improve and

The Entrepreneurial Engineer, by David E. Goldberg
Copyright © 2006 John Wiley & Sons, Inc.

innovate in creating new product lines, acquiring new customers, adopting new technology, and implementing better business practices. In larger companies, words have been coined to describe this need, *intrapreneurship* or *corporate entrepreneurship*. However, this pervasive orientation toward opportunity, innovation, and reward is now also necessary in the management of one's own career.

In times past, employers took a paternalistic view of employees, managing their remuneration, health benefits, and retirement over the course of an entire career. Those days are largely gone, and today's engineer must take charge of his or her career by seeking a challenging sequence of work experiences that help build a marketable portfolio of diverse skills. Entrepreneurial engineers meet the challenges of changing times as opportunities, seeking challenging and rewarding work together with an appropriate balance of intellectual, financial, professional, and personal growth.

1.2   ENGINEERING EDUCATION, COMMON SENSE, AND THE REAL WORLD

Common sense and general education prepare today's engineers for some of the challenges of our fast-paced times, but the predominant emphasis of an engineering education on the technical side of the ledger is, in one sense, misleading. The average engineer studies many long years, plunges into the real world, and finds that a tough part of the job was left as an exercise to the reader. While engineering school spends a majority of the time on difficult technical subjects—science, mathematics, and engineering—these important topics may constitute less than half of an engineer's working day. Moreover, career success as an engineer, while tied to technical prowess, may have as much to do with your ability to communicate with co-workers, sell your ideas, and manage your time, yourself, and others. These crucial nontechnical skills will determine your career success—and your happiness—more often than will your ability to manipulate a Laplace transform, code a Java object, or analyze a statically indeterminate structure.

Of course, this is not to belittle technical skills; after all, engineers wouldn't be engineers unless they knew a great deal about technical matters. Engineering education must spend a preponderance of its time on technical matters to bootstrap the engineer into a world of increasingly complex and changing technology. On the other hand, preparing for difficult organizational and people-related challenges helps engineers to be more effective throughout their careers.

Another side effect of the necessary concentration on technical subjects in an engineering education is that sometimes engineers think of nontechnical subjects as soft or easy—mere common sense. But the human and business sides of engineering are extraordinarily challenging and have the potential to be extraordinarily rewarding. Although many of the key topics may be classified under the category of common sense, actually putting them into practice consistently and regularly requires practice and hard work.

1.3   TEN COMPETENCIES FOR THE ENTREPRENEURIAL ENGINEER

So perhaps we agree that an engineer needs to be skilled in matters nontechnical. But what areas are particularly important to understand? Here we proceed from matters of individual concern, to interpersonal matters, to matters of team and organizational significance. Specifically, an entrepreneurial engineer should:

1. Seek the joy of engineering.
2. Examine personal motivation and set goals.
3. Master time and space.
4. Write fast, revise well, and practice BPR (the elements of background, purpose, and road map).
5. Prepare and deliver effective presentations.
6. Understand and practice good human relations.
7. Act ethically in matters large, small, and engineering.
8. Master the pervasive team.
9. Understand leadership, culture, and the organization of organizations.
10. Assess technology opportunities.

The importance of each of these is briefly reviewed in what follows.

Seeking the Joy of Engineering   The terms *joy* and *engineering* aren't often used in the same sentence, which is a shame because a proper understanding of engineering leads us to understand how multifaceted the learning and practice of engineering can be. Some of the confusion is the result of two historical inversions in perspective, and another portion of misunderstanding comes because engineering is wedged between business and science. A closer reading of history and understanding the fundamental tug-of-war that engineers face help us understand the essence and joy of being an engineer more deeply.

Examining Personal Motivation and Goal Setting   Understanding what motivates a person in his or her professional life is fundamentally important and difficult. Many people think that they simply work for the money, and indeed financial remuneration can be a factor in career choices, but a more reliable guide to a life of fulfilling work is found in the term *engagement*. Instead of seeking money directly, another approach is to seek work that is so engrossing that time flies because it is so much fun. Incidentally, the fun of engagement can lead to sufficient time on task and professional growth that the person also gains a substantial income along the way.

Mastering Time and Space   Understanding personal motivation and setting goals may be thought of as the strategic level of managing yourself. Time and storage management is the tactical level of being personally organized. How one spends one's day and where one puts one's stuff are basic to sustaining a high level of personal productivity, but a haphazard approach to these matters is all too common. Fortunately, significant improvements in time utilization can be achieved through the development of a few key habits. Primary among these are the disciplined use of (1) a calendar, (2) a to-do list, and (3) a systematic filing system (both paper and electronic).

Writing Fast, Revising Well, and Practicing BPR   Writing is as popular with many engineers as going to the dentist, but entrepreneurial engineers spend a fair amount of their business days writing. With a cubicle piled high with writing projects, the last thing the entrepreneurial engineer needs is a case of writer's block, but common writing maladies can be traced to difficulties in writing process, content, or both. Writing process can be improved by separating writing from revision, and we examine a number of specific techniques including freewriting, quick planning, and cut-and-paste revision to help us separate these two writing functions. Writing content can be improved by understanding three key elements that are common to almost all business writing. These elements—background, purpose, and road map, or BPR for short—can and should be iterated at different levels of a document.

Preparing and Delivering Effective Presentations   PowerPoint presentations are now a way of business life, but giving a slide presentation is different than giving a speech. By preparing and treating Powerpoint slides as note cards we share with our audience, the process of preparing and delivering a presentation is simplified. The rules of presentation organization are remarkably similar to the BPR rule of effective writing, and indeed good writing leads to good presenting and vice versa. Add some guidelines and concern for effective slide layout and presentation delivery and the entrepreneurial engineer is well on the way to becoming an effective presenter.

Practicing Effective Human Relations   Formal communication skills such as writing and presenting are a good place to start in developing interpersonal skill, but effective human relations are especially important in an increasingly interconnected world. The world of people can seem a lot less predictable to engineers accustomed to Newtonian models of physical systems, but engineers can find modeling guidance—and success in their relationships—by using a variation on the golden rule we call the other-eyes principle. This principle recommends that we predict or anticipate the behavior of others by considering our own reaction to a similar set of circumstances. Although not infallible, such modeling is often a good first-order guide to predicting the response of others. Along the way, we consider the importance of questions, salesmanship, praise, and passion in successful interpersonal relations. The downside of criticism is visited as is the need for admission of wrongdoing and apology.

Acting Ethically in Matters Small, Large, and Engineering   In some ways, the whole topic of ethics may be viewed as a logical extension of human relations, and the golden rule in both of its major forms is a useful entry point to the topic. Thereafter it is useful to consider a number of sources of ethical thought and reasoning, including religious or cultural norms, an innate moral sense, maximization of societal utility, and consistency. Each of these viewpoints is helpful, and each has been challenged by ethical skeptics over the years.

In moving from ethical theory to ethical practice, a key question is why people who accept a set of moral principles fail to do the right thing, and self-interest, obedience to authority, and conformity to the group are highlighted as major culprits. With this in mind, we suggest that practice on small matters is the surest way to doing the right thing when the big issues arise. In other words, our approach seeks success in *microethics* or ethical behavior in small, everyday matters. If we are unable to behave ethically when the stakes are small, it seems unlikely that we will be able to behave ethically when the rewards of unethical behavior are great.

This logically leads to a fuller consideration of professional ethics and the larger *macroethical* challenges of being an engineer. Our examination of engineering ethics concludes by considering (a) what a profession is and (b) two contrasting engineering codes of ethics.

Mastering the Pervasive Team   Teamwork has become integrated into the fabric of modern organizational life as a result of the quality revolution, but effective teamwork is difficult as many of us know from our early experiences with "group projects" in school. In group work, it becomes difficult to coordinate inter-related pieces of a project, and individuals can sometimes be uncooperative or even shirk their responsibilities. A clearheaded approach to teamwork acknowledges these difficulties and then designs team rules, discussion protocols, and other procedures to facilitate effective meetings and team activity. Our approach combines quantitative and qualitative models in a quest to create more effective teams.

Understanding the Leadership, Culture, and Organization of Organizations   Managing ourselves effectively and working well on teams are important, but understanding organizations and leadership at a somewhat higher level is also important to the entrepreneurial engineer for two reasons. First, knowing good organizations and leadership helps us pick the best work opportunities. Second, an orientation toward opportunity often results in the need to lead an existing or new organization.

A common feature of good organizations is that they think good thoughts about their employees; bad organizations tend to distrust theirs. This dichotomy is well reflected in modern theories of organizational behavior. Good companies and good leaders have been studied empirically on a comparative basis, and a number of recent studies are examined with an eye toward using their practical suggestions to better understand good companies, cultures, and leaders.

To understand organizations it is also helpful to know the economics behind their formation, and this leads to the topic of *transaction costs*. This in turn leads to a discussion of the growing trend toward more frequent job changing and free agency, and this leads us to think in terms of our network and our portfolio of accomplishment.

Assessing New Technology Ventures The entrepreneurial engineer, whether in a startup or part of a larger corporation, is increasingly being asked to help assess and pursue technology opportunities. Technology opportunity assessment and planning are challenging activities that require the entrepreneurial engineer to imagine new opportunities, match them to markets, and determine whether they are both technologically and financially feasible. As an educational experience, working on a technology opportunity assessment or a business plan is the quickest way for the entrepreneurial engineer to bootstrap him- or herself into understanding the importance of sustainable competitive advantage, customers, marketing, pricing, costs, and value at the core of the business side of engineering.

These 10 competencies are at the heart of what entrepreneurial engineers need to master to be effective in a faster moving world of deals, teams, startups, and innovating corporations.

1.4 THREE PRINCIPLES

The entrepreneurial engineer has many nontechnical skills to master, but at their root, these manifold skills of managing time, space, people, and money can be reduced, in many cases, to just three recurring principles:

1. Seek engagement.
2. Create first; criticize later.
3. Analyze through the eyes of others.

In the remainder of this section, each of these is discussed briefly in turn.

Seek Engagement We have already dealt briefly with *engagement*—finding and doing things you enjoy profoundly—as a key skill to understanding personal motivation. Fortunately, an engineering education provides a broad platform of technological and general knowledge from which to seek engagement, and the profession itself is sufficiently multifaceted that it can comfortably accommodate a wide range of motivations and personal preferences.

Create First, Criticize Later A second principle that reverberates through the 10 core competencies is that of *creating first, criticizing later*. In any creative activity, whether it be designing, writing, presenting, or problem solving, it is important to get many ideas out on the table to permit their cross fertilization and to stimulate additional associations. School learning, with its emphasis on *convergent thinking*—applying what has already been discovered—discourages the *divergent* or *lateral* thought required for excellence in all creative activities. Writing and brainstorming are fairly obvious activities where this principle applies, but the principle is general in its scope of application, and should be kept in mind in all creative endeavors.

Analyze and Design Behavior through the Eyes of Others The last theme running through the 10 competencies is *analyzing and designing behavior through the eyes of others*. The economist sometimes has the luxury of imagining a "Robinson Crusoe" economy where a single individual's wants and needs are met by his or her own efforts, but almost all human activity is carried out through a messy mix of cooperation and conflict with others. Efficiency requires that we maximize the former and minimize the latter; to do this, we must understand the motivations of the other individuals involved. It is interesting that we are pushed in this direction not by altruism but by pursuit of personal effectiveness. It is also interesting that our study of others often results in our becoming better—marginally more objective—observers of our own behavior.

Although it is impossible to capture the complexity of all human interaction in a few short words, learning and using the three principles can often lead to useful guidance in new or unanticipated situations.

1.5 THREE CAUTIONS

Engineering teaches us to apply models in an almost promiscuous fashion. Engineers model structures. Engineers model circuits and control systems. Engineers model manufacturing workstations, assembly lines, even whole factories. But when it comes to modeling creative activity and human interaction, we need to be somewhat more cautious in our modeling. Specifically, three cautions must be exercised.

Be Realistic in the Application of Ideals Many nontechnical skills are discussed in terms of *ideals*. Doing this has benefits and risks. The primary benefit of using ideals is that we can easily define a target by which we can measure our own behavior and make adjustments. The primary risk is that no one can live up to ideals all the time. My introduction to the literature of success, which came during my first job, led to frustration; I read many books about business and management and saw that my company and my clients' organizations were far from the ideals discussed. Focusing on these discrepancies led to unproductive rounds of playing "ain't it awful." This is a rookie's mistake.

One should be sensibly realistic in applying ideals. It may be all right to press for your own top performance or that of an organization that reports to you, but applying ideals to others who do not share your vision is a prescription for unhappiness and disappointment. Moreover, as pointed out elsewhere (Fritz, 1991), you should select ideals with considerable care, as it is possible to become paralyzed by conflict between what the ideals promise and what is actually possible.

Mastering the Obvious Isn't Easy Engineering students often think of practical nontechnical matters as being "common sense" or "obvious," but just because some skill is superficially obvious doesn't mean that mastering that skill is easy. The problem here is the inherent complexity of many nontechnical skills. For example, writing is a many-layered topic. At root, writing is about such basic skills as grammar and punctuation. At the next level, it is about sentence structure, flow, and paragraphs. At yet another level, it is about overall organization and presentation of material in a logical yet interesting fashion. With so many levels requiring adequate performance, it is difficult to be proficient on all of them simultaneously, and there is always room for improvement on one of them.

In talking about another complex skill—swinging a golf club—champion golfer Ben Hogan wrote the following (Hogan, 1957, p. 30):

> It may be seen that we have gone into unwarranted detail about the elements of the correct grip [of a golf club]. This is anything but the case. Too often in golf, players mistake the generality for the detail. They think, for example, that overlapping the finger is the detail and so they do not pay sufficient attention to how they do it. Or they confuse an effect (which can be quite superficial) with the action (the real thing) that causes the effect.

Like golf, the core nontechnical competencies required of the successful entrepreneurial engineer are sufficiently complex that there are many subtle details or facets to learn; it is impossible to learn them in a single lesson. Moreover, it takes conscientious practice to maintain a skill once it has been developed. Just as golf professionals return over and over to the fundamentals of grip, stance, swing plane, and so on, so too must we return repeatedly to the basic skills and principles that help make us effective in our professional lives. And just as professional athletes take a positive attitude toward continual improvement—a philosophy the Japanese call *kaizen*—it is especially important for entrepreneurial engineers to adopt a philosophy of continuing professional development.

Engage the Material and Put It into Practice   Discuss a technical problem with data, equations, and graphs and a group of engineers will become engaged and animated. Discuss a personnel problem with interpersonal drama, tension, and ambiguity and that same group's eyes will glaze over. And unfortunately the personnel problems are getting in the way of our doing the fun technical stuff as much and as well as we would like.

No one will agree with this text on all matters, but when you find an idea, suggestion, or tip that makes sense, why not try applying it to your work? When you disagree with a topic or approach, why not read what others have to say about the same subject? The bottom line of this book is that there is more to engineering than technical skill, and engaging the material herein, questioning it, and putting it to practice is a good start toward becoming a more entrepreneurial engineer.

EXERCISES

1. Make a list of the five people who have influenced your life most directly and the ways in which they changed the course of your life to date.
2. Make a list of between one and five living public figures you look to as role models. What elements of their character do you wish to emulate. List at least one specific experience or event about each figure that you particularly admired.

3. Make a list of between one and five living public figures you least admire. What elements of their character do you find objectionable. List at least one specific experience or event about each figure that exemplifies your concern.

4. Read a bibliography of an historical figure you admire. Rate the person on his or her (1) character, (2) interpersonal dealings, (3) communications ability, (4) organizational capability, (5) balance between family and work, (6) technical skills, and (7) leadership ability. List a single anecdote from the person's life that is most telling of his or her nature as a person.

5. List the reasons why you became (are becoming) an engineer. If you had the decision to do over, would you make the same decision? Why or why not? In what ways has the decision turned out better than you expected. In what ways have you been disappointed by the decision.

6. What experiences to this point in your life have most influenced your ability to work with others? Make a list of those experiences and list the primary lesson or lessons of each experience.

7. Make a list of your top three strengths. List the ways in which those strengths connect to your career or life decisions to date. List five ways in which you could build on each of those strengths.

8. Make a list of your top three weaknesses. List the ways in which those weaknesses have limited your career or life progress. List five ways in which you could help overcome each of those weaknesses.

9. Make a list of three companies or organizations you admire. Explain your choices in a short essay.

10. Make a list of three companies or organizations you do not admire. Explain your choices in a short essay.

Chapter 2

The Joy of Engineering

2.1 A JOYOUS CONFESSION

I have a confession to make. I am an unabashed, card-carrying engineering *chauvinist*. I believe that engineering is a terrific education and that engineering practice can be a joyful way to spend one's life. In a technological age, engineers are constant, yet largely unsung, contributors to our quality of life, the creators of the systems, processes, and products that we depend upon day to day. Every time you step on or off a modern airplane, you literally owe your life to the hard work of teams of anonymous engineers. Every time you bang out a memo or essay on your laptop computer, you owe a hearty thanks to thousands of unheralded hardware and software engineers. Everyday you wake up, make breakfast, go to work and return, you trust your safety to the handiwork of myriad infrastructure engineers who safeguard the water you drink, the highways you drive, and the electric power grid into which you plug.

Yet our culture hardly gives these engineering contributions a second thought. TV celebrates lawyers, doctors, businessmen, politicians, and even forensic scientists, but engineers are merely the steady Freddies and Janes that build, operate, and maintain the stuff we use. Ironically, it is in this sense that engineering is a victim of its own success. We *can* depend on the airplanes, the computers, the software, and the infrastructure, so there is little to dramatize in prime time. If only we engineers, as a group of professionals, would mess up more often, perhaps the foibles of flawed engineering practice would be sufficiently dramatic for prime-time TV. Yet, it doesn't seem quite right to be so blasé about a discipline and a group of people who are arguably so important to modern society's day-to-day safety and success. Moreover, our culture's lack of attention to the artifacts and people of engineering causes it to misunderstand engineering education, engineering practice, and engineers themselves in important ways.

This chapter critically examines an engineer's place in the world. Although it may continue to be difficult to convince nonengineers of the importance of what engineers study and do, it is important for engineers themselves to have a better understanding of key historical, philosophical, and methodological foundations of their discipline and profession. We start by viewing an engineering education as a new kind of liberal education and as a practical career launchpad. Thereafter we consider the role of engineering and engineers in the establishment of modern business practice. This leads to the incorporation of some economics in a discussion of the methodology of engineering practice and a discussion of four intellectual tensions faced by the postmodern entrepreneurial engineer.

2.2 ENGINEERING AS LIBERAL EDUCATION, LAUNCHPAD, AND LIFELONG LOVE

Not long ago, engineering education and working as an engineer were viewed as reliable ways to get a leg up into the middle class. Today, engineering education is a launchpad for a variety of careers. Moreover, it is increasingly being viewed as a broad education appropriate to a time of increasing technological sophistication, and this breadth leads today's savvy engineering graduates to enjoy the fruits of their education and careers in a variety of ways. In this section, we analyze the breadth of an engineering education by comparing the requirements of an engineering degree with those of an English degree. We examine some of the different careers that an engineering education can launch, and we consider a number of ways in which engineering is an engaging field of study and work.

2.2.1   Who Is Getting a "Liberal Arts" Education Today?

The notion of a broad liberal arts education goes back to the Greeks and Romans and has been at the center of the modern secular university. Engineering education has a somewhat shorter history, but the perception that engineering education is extremely narrow by comparison misses several key points:

- The modern engineering curriculum is remarkably balanced and substantially in line with key elements of a classical "liberal education."
- The centroid of knowledge has shifted toward matters technical to the point where a classical liberal education with little math or science emphasis and no study of technological artifacts is no longer a broad basis for understanding the world around us.

The first point to make is that engineering education can be remarkably balanced. Consider the breakdown of the general engineering curriculum at the University of Illinois at Urbana–Champaign shown in Table 2.1. The first 3 rows in the table total to 79 hours or 60 percent of the total. Interestingly, those hours would satisfy core or distribution requirements in almost every liberal arts major. The 22 hours under mechanics and engineering science specialize math and science topics to engineering practice and treat topics that upper-level undergraduate majors in science or math would cover. Only the 30 hours in the major are truly specialized and could not fit under the liberal arts rubric. Thus, only 23 percent of the 131 hours can be thought of as truly specialized, and that number

Table 2.1 Breakdown of an Undergraduate Engineering Education

Category	Semester hours
Humanities and social science	25
Secondary field (tech or nontech) + electives	18
Math and science	36
Mechanics and engineering science	22
Major design and analysis specialization	30
Total	131

is comparable to the number of hours that a liberal arts student would take in his or her own major.

If we turn this around and examine the composition of the requirements of a typical liberal arts degree, for example, an English degree, we notice a number of things (Table 2.2). First compare the total hours. Using the University of Illinois again as an example, an English major can graduate with 11 fewer hours than the general engineering major. Engineering curricula commonly require roughly one-half to three-quarters of a semester more to graduate than a comparable liberal arts degree. In other words, some of the specialization of an engineering degree is paid for by working longer.

But a closer look at the distribution of hours is even more interesting. Required social science and humanities hours outside the major are comparable in the two degrees; however, the English degree requirement of only 9 hours of math or science is striking. It hardly seems reasonable that a broadly educated person in an age of rapid technological advance should be able to get a bachelors degree with only 9 hours, or 7.5 percent of the total hours in the degree, studying math or science. Can a person claim to be broadly educated with such a paltry number of hours in exactly those subjects that are advancing most quickly? No, the centroid of knowledge has shifted—and continues to shift—toward science and mathematics, and the "liberal arts" degree of times past does not—in and of

Table 2.2 Breakdown of an Undergraduate English Major

Category	Semester hours
Humanities and social science	24
Minor subjects	18–21
Math and science	9
Electives	36–39
English specialization	30
Total	120

itself—serve to broadly educate those who receive it. Add in the loosening of rigorous core distribution requirements and core courses that has taken place since the 1960s and 1970s, and it is not a stretch to argue that the kind of rigorous engineering degree such as that profiled above is a broader, more appropriate degree for educating well-rounded people in an increasingly technologically intensive world.

2.2.2 Engineering as Launchpad

In times past, an engineering degree was largely expected to prepare individuals for a professional career working as an engineer. Although many engineering graduates continue to find employment as working engineers, an engineering degree is also being viewed today as good preparation for other careers. Large numbers of engineering students apply to medical school, dental school, law school, and business school following their undergraduate engineering education.

Some older engineering faculty members find this trend disconcerting and wish for the good old days when engineering students graduated and worked as engineers. Yet the broader acceptance of engineering education—and values—by those who don't work as engineers can be viewed as a blessing. First, a more technologically educated populace will better appreciate technology and the challenges of its care and feeding. Moreover, medical and legal professionals as well as high-level managers are often influential members of a community, and having such people both knowledgeable in and sympathetic to engineering and technological matters should be beneficial. Finally, discussions of increasing engineering influence often lead to suggestions about having engineers lobby government, Hollywood, or the media. But perhaps the more powerful kind of influence occurs when engineering students take their engineering education and win seats in legislatures, earn positions in movie and TV production, or find positions as working journalists. None of this is to recommend those particular career paths for those who are not interested in them, but the key point is that an engineering education is a broad one that can prepare its recipients for careers across the spectrum of human endeavor.

2.2.3 Ten Ways to Love Engineering

When I ask engineering students to tell me why they decided to come to engineering school, some will talk to me about their hobbies or interests (cars, radios, computers, software) and their eyes light up, and I'm reasonably confident that I'm talking to someone who will make engineering school work for them. When students start talking to me about high school guidance counselors, good grades in math and science, or high pay from engineering jobs, I get a little nervous that I'm talking to someone who hasn't found something to love about engineering yet. Engineers use math and science and like to think of themselves as fairly rational beings, but the engineers who succeed longest and best are those who have found something to love about their chosen path.

Fortunately engineering is multifaceted and we can find joy in engineering because it is

1. Creative
2. Intellectually stimulating and challenging
3. Concerned with the real world
4. Constructive
5. A people profession
6. A maverick's profession
7. Global
8. Entrepreneurial
9. Optimistic
10. An entry point to lifelong learning

First and foremost, ours is a creative profession. As engineers we create that which has never before existed, through a combination of imagination, ingenuity, and perseverance. We therefore have many opportunities to become engaged in the creative processes of idea generation and problem solving. This stands in stark contrast to those professions that train their practitioners largely to become proficient in extant technology and technique.

Engineering is also intellectually stimulating and challenging. Being a good engineer requires much knowledge and know-how, but no armchair intellectuals need apply. Ours is a profession that requires streetwise application of mind to means, where the touchstone of success is whether the job gets done.

This leads us to recognize that engineering is firmly rooted in the real world. This has a number of benefits. It forces us to face up to the limitations in our modeling, and it forces us to confront difficult variables that defy analysis—variables such as time, money, consumer preferences, the impact of government, and the impact of technology on society.

Moreover, engineering is an inherently constructive profession, attempting to make a better world through change. Contrast this to some other professions that add costs and paperwork to many transactions without adding direct value to the processes involved or products produced. Engineers often find great pleasure in being able to touch or see the results of their labors, taking great pride in their contribution to a completed product or project.

Ours is a people profession as engineers often work in teams. As marketing, manufacturing, and engineering considerations are integrated into the design process, engineers increasingly find themselves working on teams with many different types of individuals across a company. Of course, we've devoted a good bit of space in this book to emphasizing the habits necessary for good interpersonal relations, whatever the circumstances; but the engineer who is skilled in his or her dealings with others will also find many opportunities for engagement therein.

At the same time that engineering requires team effort, it can also call for outstanding individual effort. Many of the most creative and advanced engineering projects have required a champion to almost single-handedly overcome obstacles and single-mindedly bring an idea to fruition. Thus ours is a profession that finds a place for the engaged maverick at the same time it embraces the team and its players.

Our world has become a very small place. Jet travel allows us to become physically present almost anywhere in the world in less than a day. Satellites, fiber optics, and advanced computing allow us to become virtually present almost anywhere in the world in milliseconds. Such changes are making engineering a more global profession, where products are designed and built across borders, even across oceans. This situation creates opportunities for the engineer who is willing to learn other languages, customs, and cultures.

Some of the same technological influences that make engineering a more global profession are opening up new entrepreneurial opportunities. Engineering has had a long tradition of private practice and private enterprise, but the tumult in such fields as modern electronics, information technology, biotechnology, and nanotechnology has opened new vistas for the engineer-entrepreneur. As communications technology makes close ties at a distance a reality, more and more engineering functions will be farmed out to independent design shops at remote locations. At the same time, the tools of our trade have dropped in price; the small shop need be at little or no competitive disadvantage to the in-house engineering operation of a Fortune 500 firm. Moreover, miniaturization, machine-tool, and materials-handling technology is driving manufacturing toward point of sale. As we move in such directions, it should be clear that a company's competitive advantage will lie more in its intellectual property—in its designs—and less in its manufacturing and distribution capability. Although the present has been kind to the entrepreneur-engineer, the future holds many engaging opportunities for those with the enterprising spirit.

It almost goes without saying that ours is an optimistic profession. Our impulse as builders is reinforced by the knowledge that we have improved what was once a very hard life and the hope that our continued efforts can make things even better. Sometimes we have paid insufficient attention to the unintended consequences of our acts, but the genie of innovation prefers freedom to the confinement of the bottle, and once free he has largely served his masters well.

Finally, as pointed our earlier, even for those who get an engineering degree but choose not to practice as engineers, engineering education positions its recipients for lifelong learning and growing. Given the conversance with mathematics, science, technology, the humanities, and the social sciences required in a Bachelor's degree in engineering, those with an engineering education can pick up texts in almost any subject and learn. The broadly educated engineer of today is thus better positioned for lifelong learning and growth than those who have not struggled with the artifacts, knowledge, language, or details of technology.

Thus, the ways to find fulfillment through engineering are many, but the very breadth of the engineer's purview gives rise to a fundamental tension inherent in both learning and practicing engineering. This tug-of-war is our next concern.

2.3   THE FUNDAMENTAL TUG-OF-WAR

Engineering is an old and venerable practice, but in modern times engineers oftentimes find themselves in something of a professional vise. On the one hand, they find their work lives ruled by managers, accountants, and other business school graduates. On the other, they find their profession itself criticized by scientists and mathematicians as being the *mere* application of science and math to problems of practical import. Since its inception as a modern profession, engineering has been a combination of commerce and science (Layton, 1990). But the vise hold of these two disciplines undervalues the creativity of engineering analysis and design (Vincenti, 1990), and it misjudges the delicacy of the hybrid of analytical and interpersonal talents engineers must master to be successful in practice. Moreover, the squeeze play hides the historical record of engineering's role in the formation of both science and engineering.

We start by considering two historical inversions—one between science and engineering and one between business and engineering—that have permitted engineering, the field that arguably has historical priority in both cases, to be caught in the middle, both commercially and intellectually. We distinguish between the commercial and scientific aspects of the engineering mind using an economic model of the modeling process, and consider the spectrum of models from qualitative and quantitative implied by the economic model. This in turn leads to a discussion of four tensions facing the entrepreneurial engineer.

2.4   SCIENCE AND ITS LITTLE SECRET

I once was having a discussion with a colleague in physics from a major research university at a meeting sponsored by the National Academy of Science. He offered his opinion that engineering is "just applied science, nothing more, nothing less," and this sounds plausible enough to modern—even engineering—ears. Today's engineering education dwells on math and science first, and engineering subjects are taught as practical elaboration or embellishment of those more primary subjects. It wasn't always this way. If we return to the origins of modern science itself, we understand how the engineers of that time were inspiration for the remaking of the enterprise we now recognize as science. In particular, we review how the man often called "the father of modern science," Sir Francis Bacon, used 17th-century engineering practice as inspiration for his reformulation of the scientific method (Figure 2.1).

At the time, natural philosophy was stuck, and Bacon tackled the problem in his book *The Great Instauration*. His first task was to acknowledge the problem (Bacon, 1620/1994, p. 6):

Figure 2.1 Francis Bacon (1561–1626) (reproduced with permission of Maxwell C. Goldberg).

That the state of knowledge is not prosperous nor greatly advancing; and that a way must be opened for the human understanding entirely different from any hitherto known ... in order that the mind may exercise over the nature of things the authority which properly belongs to it.

Interestingly, Bacon's motivation was in large part that of the engineer. The need for better science was so "the mind may exercise over the nature of things the authority which properly belongs to it."

His next job was to analyze why knowledge was not advancing as fast as it might. He laid blame at the feet of the vast majority of philosophers of his time who blindly believed in the ancient dictates of Aristotelian physics (p. 7):

Observe also, that if sciences of this kind had any life in them, that could never have come to pass which has been the case now for many ages—that they stand almost at a stay ... and all the tradition and succession of schools is still a succession of masters and scholars, not of inventors and those who bring to further perfection the things invented.

Not only did Bacon question continued blind allegiance to the masters of the ideas of Greece, he foreshadowed his solution for recasting science with his curious choice of the words *inventors* and *things invented*. This led readily to a call for recasting the methods of philosophy along the lines of 16- and 17-century engineering practice (pp. 7–8):

> *In the mechanical arts we do not find it so: they, on the contrary, as having in*
> *them some breath of life, are continually growing and becoming more perfect.*
> *As originally invented they are commonly rude, clumsy, and shapeless;*
> *afterwards they acquire new powers and more commodious arrangements and*
> *constructions; ... Philosophy and the intellectual sciences, on the contrary,*
> *stand like statues.*

Bacon lived during a time of increased technological improvement. For him, the contrast of rapid technological evolution with the dearth of progress in knowledge was almost unbearable. Moreover, his solution to this critical problem was clear. Philosophy must adopt the attitudes and methods of the "mechanical arts" in the invention of concepts. Quite clearly, in Bacon's time, engineering was not merely "applied science." Far from it. Bacon's grand plan for recasting science, systematizing scientific method, and advancing the state of knowledge was inspired directly by the application of engineering method to the invention of new concepts!

Thus, it is more than a little interesting that engineering, arguably the master discipline that showed the way out of the dead end of Aristotelian physics, has become subservient to the sciences it inspired. Indeed it is entirely proper that modern science should help drive the advance of modern engineering just as modern engineering helps drive the advance of modern science. But there is no historical basis for an engineering inferiority complex (physics envy) or related maladies of modern times. If anything, to those scientists who would say that engineering is the "mere application of science" it is entirely historically accurate—if equally haughty—to reply that science is the mere application of engineering method to the invention of concepts.

2.5   ENGINEERS: FIRST MASTERS OF MODERN ENTERPRISE

Bacon used the mechanical arts (engineering) to inspire the reformulation of natural philosophy, but modern memories are short and tend to think of science as the master discipline. An analogous historical inversion has occurred in business. In business today, enterprise is run by a professional class of trained managers, and engineers are viewed merely as one instrument of a larger capitalist enterprise. It wasn't always this way.

Prior to the 1850s, business was performed on a relatively small scale. What large-scale businesses there were could be organized along fairly decentralized lines, thereby requiring methods that were no more complex or coordinated than those of smaller enterprises. All that changed with the coming of the railroads (Chandler, 1977, p. 80):

> *Of the new forms of transportation the railroads were the most numerous, their activities the most complex, and their influence the most pervasive. They were the pioneers in the management of modern business enterprise.*

Because of their historical importance, Chandler is careful to identify these pioneers of modern business (p. 87):

> *The men who managed these enterprises became the first group of modern business administrators in the United States . . .*

To a man, they were engineers (p. 95):

> *The men who face these challenges were a new type of businessman. . . . The pioneers of modern management . . . were all trained civil engineers with experience in railroad construction and bridge building before they took over the management of the roads.*

Many of these same people had military training, and Chandler wondered whether the new business methods they developed were borrowed from the management of men and material in the military. Chandler rejects this hypothesis emphatically (p. 95):

> *Yet even for such officers, engineering training was probably more important than an acquaintance with bureaucratic procedures. There is little evidence that railroad managers copied military procedures. Instead all evidence indicates that their answers came in response to immediate and pressing operational problems requiring the organization of men and machinery. They responded in much the same rational, analytical way as the solved mechanical problems of building a bridge or laying down a railroad.*

Thus, in the early days of modern enterprise, engineers were the innovators who developed the methods—the profession—of modern business. As with the scientific inversion, here there is no historical basis for "business envy" or a "nerd inferiority complex." The businessperson who says that engineering is "mere technology applied to the needs of business" could more accurately be told that modern business is merely the application of engineering method to the design of commerce.

2.6 ECONOMY OF INTELLECTION: SEPARATING SCIENCE FROM ENGINEERING

Historical analysis helps shed some light on the professional vise grip engineers find themselves in, but modeling how engineers are different from scientists and businessmen is a more difficult matter. Here we distinguish engineers from scientists in terms of their *use of models* through an argument based on an *economy of intellection* (Goldberg, 2002).

Scientists and engineers both build and use models of physical phenomena—often mathematical models—but the motives and economics behind their model usage are distinct. Engineers create and use models to advance technology, whereas scientists primarily build ever more accurate models of observed phenomena.

2.6.1 Modeling Plane

All of these models live on a plane of error and cost (Figure 2.2). Engineers and inventors use models of relatively high error and low cost, whereas scientists and mathematicians build and use models of relatively low error and high cost. All this makes sense when viewed in the light of the distinct objectives of engineers and scientists. Scientists are in the business of reducing the error of current best models and should be expected to spend most of their time at the high C, low ε portion of the curve, pushing for lower and lower ε regardless the C.

The engineer's position is a little harder to understand and justify, but straightforward principles of economics come to the rescue. Imagine an engineer faced with the prospects of going from a model of error ε_1 to one of ε_2. The move incurs a marginal cost $\Delta C = C_1 - C_2$, but unlike the scientist, the engineer is generally not in the business of building better models for their own sake. No, the engineer is usually charged with improving some *technology*—some product, service, or process—and ostensibly the use of an improved model should yield some *benefit* to the technology of interest. In mathematical terms, there should be some marginal benefit, $\Delta B = \Delta B(\varepsilon_1, \varepsilon_2)$ to the technology that results from the use of a more accurate model. In practical terms, this benefit can come from better qualitative or quantitative understanding of the mechanisms underlying the technology, but for the engineer to justify the use of a higher cost model, some improvement should be expected. Moreover, if the benefit can be stated in monetary terms, the engineer can be said to be engaging in *economic* model

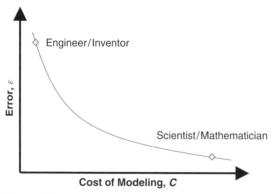

Figure 2.2 Models live on a plane of error and cost. Engineers generally use lower cost, higher error models to benefit some object technology. Scientists and mathematicians are usually more interested in the creation of new models. [Adapted with permission from Goldberg (2002).]

selection if the marginal benefit of the model to the technology equals or exceeds the marginal cost of the improved model:

$$\Delta B \geq \Delta C \qquad (2.1)$$

If the engineer chooses the more expensive model when the above condition does not hold, the decision is said to be *uneconomic*, and at least some amount of the improved accuracy of the model is wasted.

Of course, none of the foregoing discussion should be taken to suggest that engineers actually perform the formal cost–benefit analysis as part of their modeling usage. The costs of modeling are not often explicitly calculated, and even if they were, the benefits of modeling are difficult to quantify and usually unknown in advance of the engineering design effort. Nonetheless, modern engineering education implicitly recognizes economy of thought in the engineering curriculum. For example, where physics courses teach Newton's second law first ($\Sigma F = ma$), the engineering curriculum teaches statics ($\Sigma F = 0$) before the second law. Spending a full semester on tedious equilibrium problems seems like a waste of time and effort to most physics professors, but by doing so, the engineering curriculum drives home the important lesson of *grabbing the cheaper model first*. Engineering in this sense is distinct from science, and the use of less precise models in the engineering process is economically essential; to do otherwise would be foolhardy, irrational, or both. Interestingly, the respect for economics that separates engineer from scientist is exactly what ties engineer to businessperson.

2.6.2 Spectrum of Models

Cost–error analysis of models suggests a one-dimensional spectrum of models from fairly high-error (low-cost) models to low-error (high-cost) models as shown in Figure 2.3. On the far left, we have wisdom that is known but difficult to articulate (unarticulated wisdom or tacit knowledge). Moving to the right we have qualitative knowledge articulated in words; much business knowledge is of this sort as is knowledge of history and many of the humanities. On the far right we have equations of motion that specify the trajectory of a dynamic system in some reasonably complete manner, and to the left we have so-called *facetwise models* (Goldberg, 2002) in which various simplifications are made to equations of motion or their solution to obtain a model of a single facet of a more complex motion system. In the middle of the spectrum we have the entry point into quantitative modeling using dimensional analysis and scaling laws.

At any given point in one's work life, engineers will be called on to use a combination of tacit knowledge, articulated qualitative knowledge, dimensional reasoning, facetwise models, and full equations of motion. Although the agility and breadth of modeling skill required is difficult to learn, it is essential that it be mastered.

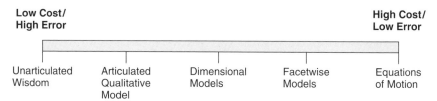

Low Cost/   **High Cost/**
High Error   **Low Error**

Unarticulated   Articulated   Dimensional   Facetwise   Equations
Wisdom   Qualitative   Models   Models   of Motion
                  Model

Figure 2.3   Spectrum of models goes from tacit knowledge (unarticulated wisdom) to full equations of motion. In between, qualitative and quantitative models of various degrees exist and should be mastered by competent entrepreneurial engineers. [Adapted with permission from Goldberg (2002).]

2.7   FOUR TENSIONS FACING THE ENTREPRENEURIAL ENGINEER

The modeling plane and the spectrum of models highlights a critical tension in engineering knowledge, the tension between qualitative and quantitative knowledge. Yet, the entrepreneurial engineer faces this and three other key tensions in trying to reconcile the demands of a fast-paced life driven by technology and markets:

- Qualitative versus quantitative mode of analysis
- Human-centered versus technology-centered focal point
- Centralized versus decentralized locus of control
- Mature versus immature knowledge

Each of these is briefly discussed:

Qualitative vs. Quantitative   Engineers have a cultural predisposition to speak equations and generally go quantitative, but much business knowledge is qualitative in nature. Entrepreneurial engineers need to appreciate that many topics are well handled with words rather than equations. A good way to plug into this mode of thought is to (a) read a business newspaper on a regular basis (e.g., *The Wall Street Journal*), (b) read business case studies (e.g., from the Harvard Business School), (c) read popular business books and business textbooks, and (d) read more fiction and nonfiction generally. Entrepreneurial engineers must embrace the qualitative side of their brains at the same time they attempt to go quantitative on subjects that have previously defied analytical description.

Humans vs. Technology   Another tension faced by the entrepreneurial engineer is the tension between a concern for humans and a concern for technology. Many engineers come to engineering because of their love for gizmos. I confess that I came to the profession as a card-carrying amateur radio operator who loved the smell of solder and the sound of a DX (long-distance) contact with hams (amateur radio operators) far away. Having said this, the entrepreneurial engineer lives in a world of customers, co-workers, investors, and other people, and it is important to understand that notions of truth in a technological setting are somewhat different for those of a social setting. The term *postmodernism* as used in its philosophical sense embraces notions of truth that depend on the influence of populations of people. Of course, extreme accounts that deny truth in science are untenable to scientists and engineers. Searle's account (1995) embraces science (brute facts) and social or institutional facts quite nicely and should be consulted by those seeking a useful foundation for both better understanding social and scientific knowledge. At a more practical level, human–technological conflicts will usually manifest themselves as economic or political concerns.

Centralized vs. Emergent   The Cold War engineer was an inveterate planner, working as part of a large team at the core of a single organization to bring a technical artifact to fruition. On the other hand, the entrepreneurial engineer sees his or her activity within an organization as part of a larger system, an economy, in which planning is not always the rule. Economies are themselves a messy mix of the planned and the unplanned. Markets, on the one hand, emphasize the uncoordinated emergence of competitors and commerce, while at the level of organizations, planning and centralization are the key. Current results in nonlinear, complex adaptive systems, and networks are helping scientists and engineers better understand that portion of our world that is uncoordinated, but that understanding is not yet as mature as our understanding of the world of centralized control.

Mature vs. Immature Knowledge   Entrepreneurial engineers live in a world where knowledge is required from both mature and immature disciplines. New technology oftentimes lives on the boundaries of different disciplines or the edge of the newly discovered. Either way this suggests that much of what is needed will come from immature disciplines. This would not seem to be much of a problem, except that mature disciplines have an inherent advantage over immature ones and tend to crowd out the new kids on the block. Mature disciplines are those where long years, decades, or centuries have given many researchers and pedagogues the ability to refine both knowledge and teaching sequences to the nth degree. On the other hand, the state of knowledge in immature disciplines will often be less than tidy. Methodology may be inconsistent, key questions may remain unanswered, and teaching may be poorly sequenced or spotty in coverage. Nonetheless, the entrepreneurial engineer must learn to use immature knowledge in areas of importance side by side with knowledge from old and venerable disciplines. Learning to embrace the new and the old together is challenging, but to prefer one over the other is needlessly limiting to the scope of an entrepreneurial engineer's range of influence.

SUMMARY

Engineering can be a joyful occupation, and this chapter started by examining how an engineering education is balanced, broad, and able to launch a variety of careers. This led to a discussion of 10 different ways to love engineering, ranging from finding joy in engineering as a creative and intellectually stimulating endeavor to understanding how engineering can be entrepreneurial, optimistic, and global. Because engineering is a blend of the technical and commercial, engineers often find themselves in a tug-of-war between science and commerce, and this chapter has tried to understand the engineer's position with respect to those two poles of the engineer's mind through historical analysis. Interestingly, the analysis turns conventional wisdom on its head, and engineering may be viewed as inspiration for both modern science and modern business in a historically rigorous sense.

The chapter has also attempted to bridge the gap between the scientific and commercial poles of engineering thought through an economic model of the modeling process. Although such modeling is rarely formal, engineers tacitly consider the marginal costs and benefits of the mathematical and scientific models they can apply to the development of a new technology. Balancing model rigor and cost in this way helps ensure that the costs of technological advance are well tied to the utility of the technology being advanced. Science is largely about building better models, and it is not surprising that the activity is less conscious of modeling costs. Of course, both kinds of activities are necessary, and both types of individual are doing the kind and style of work necessary to get their respective jobs done. This reasoning has also led to consider the spectrum of models from tacit knowledge to full equations of motion. The entrepreneurial engineer is advised to embrace appropriate models toward advancing opportunities in his or her enterprise.

The chapter has also considered four core tensions in the world of the entrepreneurial engineer. The tensions between qualitative and quantitative knowledge, between humans and technology, between centralized planning and emergence, and between mature and immature knowledge must be understood, then embraced. The very complexity of the world of entrepreneurial engineering demands an approach that is appropriately complex. The narrow disciplinary focus of the Cold War engineer with rigid ideas of methodology and content is unlikely to cast a broad enough net. The challenges of becoming a competent entrepreneurial engineer are great, but the rewards are commensurate to the challenge.

EXERCISES

1. Interview one or more working engineers at a large corporation and discuss with them the material of this chapter. Ask them to recall specific instances of experiencing the tug-of-war and how they dealt with it. Reflect on whether a different approach might have diffused the tension. Write a short essay discussing the interview and your reflections.

2. Consider your own dealings with business majors in college or at work. Recall whether you have experienced the historical inversion discussed. Would you now deal with such encounters differently and how so? Write a short essay discussing your recollections and reflections.

3. Consider your own dealings with mathematics or science majors in college or at work. Recall whether you have experienced the historical inversion discussed. Would you now deal with such encounters differently and how so? Write a short essay discussing your recollections and reflections.

4. Statisticians recommend that experiments be performed repeatedly to ensure that the results when averaged are statistically valid. Engineers might not always have the time or money to perform sufficient testing to satisfy the statistician's tests of statistical significance. Consider in a short essay whether it might ever be acceptable to perform

incomplete testing and relate your argument to the argument about the cost and benefits of engineering modeling.

5. In your technical specialty, give two specific examples each for (a) tacit or unarticulated knowledge, (b) qualitative articulated knowledge, (c) dimensional reasoning, (d) facetwise models, and (e) equations of motion.

6. The separation between qualitative and quantitative knowledge is not always as great as has been suggested. Give an example where qualitative reasoning and quantitative reasoning have interacted in classroom studies or work experience.

7. Read a text in an emerging scientific or engineering discipline. Characterize the ways in which the knowledge is less mature than that in longstanding fields in a short essay.

8. Modern market economies are a mix of planning and emergence, whereas communist regimes attempted to replace markets with centralized planning by the government. Write a short essay in which you consider (a) the proper balance between centralization and emergence and (b) an intellectual justification for drawing that line.

9. The modeling plane suggests that there is a marginal cost–benefit justification for the selection of a given model. Make a list of five benefits that could occur by using a more accurate model in practice. Make a list of five costs that could occur that would recommend the use of a less costly model in practice.

Chapter 3

Money, Work, and You

3.1 MONEY, MOOLA, THE BIG BUCKS

Money. Just the word has the power to cause us pause, to make us think. Many of us think our prayers would be answered and our lives would be at peace if only we had a little bit more money.

Unfortunately, it is rarely that simple; our feelings about money (or our lack of it) are often complex and sometimes mask much deeper concerns about our careers, our lives, ourselves. To be clear about this, and to set our sails to the future somewhat more confidently, we must investigate these concerns by examining the roads to wealth. A recent study of millionaires is extraordinarily revealing, and it helps us understand both the tactics and strategy of wealth creation. Some of the tactics we might dismiss as platitudes or common sense, but the strategic understanding of *vocation* and *courage* will be very helpful in formulating a sensible method for planning both the short and long term.

Now, all this talk about money, money, and money might be deemed overly materialistic if not a little bit coarse, but the endgame here is to turn the money talk on its head. I apologize if I'm engaging in a little Madison Avenue advertising hype and leading with "money" in the same way that leading with "sex" or "guaranteed weight loss" all but ensures an audience. More than a few engineers and engineering students come to the field with as much interest in material comfort as in science and technology. So to them, I lead with the "dollars" (euros, or yen), but throughout the chapter, we will work to take all this money talk to a higher plane. Indeed, once we've shifted from money to engagement and courage, it is more natural to talk about happiness, values, goals, and personal mission in life.

3.2 ROADS TO WEALTH: FOUR DINNER TABLE PLATITUDES

How do wealthy people get that way? Well, some get their money the old-fashioned way: They inherit it. Others win the lottery or are otherwise graced with good fortune, but we are less interested in these special cases than those where the fortunate have had more of a hand in amassing their millions. Those

The Entrepreneurial Engineer, by David E. Goldberg
Copyright © 2006 John Wiley & Sons, Inc.

working stiffs who find their way to financial security find many ways to get there. A study by Thomas Stanley (2001) used sophisticated geodemographic techniques to send surveys to 5063 individuals believed to be in a number of wealthy American neighborhoods. The survey had a total of 277 questions, and a total of 1001 of the 5063 surveys were returned and answered.

Stanley wanted to study people who accumulated substantial sums of wealth, not those who had substantial sums flowing through their fingers—what he called *balance sheet affluence* as opposed to *income statement affluence*. To this end, he screened the surveys and only considered those individuals with a net worth of $1 million or more. After all the data collection, processing, and screening, Stanley was left with completed surveys from 733 people. To augment these quantitative findings, Stanley also ran a number of focus groups and personal interviews to build anecdotal evidence and further probe the quantitative results.

After crunching the numbers, sifting through the interviews, and analyzing the results, what did this group of 733 people have in common that might account for their ability to amass and keep wealth? Did they share a particular line of work, such as sales or real estate? Were they particularly good at investing? What was their secret?

Stanley's first answer may disappoint you. In particular, Stanley found that substantial wealth accumulation was associated with four attributes:

1. Integrity
2. Discipline
3. Social skills
4. Hard work

I don't know about you, but when I first read this list, I thought I was listening to mom or dad give Lifeishardwork Lecture 101. These four items are so absolutely part of the conventional wisdom that they cry out for a special name. They are so close to the things repeated around the family dinner table that we will call them Stanley's *four dinner-table platitudes* (4DTPs). But just because they are platitudes, and just because you have heard them a gazillion times before, does not make them useless.

To the contrary, conventional wisdom can be right, and the 4DTPs make a lot of sense. Integrity allows others to trust you with their business, their money, and the belief you will deliver. Discipline can keep you on task through the peaks and valleys that accompany any long-term enterprise. Social skills allow you to make and keep the relationships that support your wealth-creating activity, and hard work keeps you pushing toward success.

No, the problem with the four platitudes is not that we don't believe or understand them. The problem is that it can be so difficult to practice them well day in and day out. There must be other factors that differentiate successful people from the rest of us. What keeps them working hard with discipline while building their networks and being as true as honest Abe? The key to cracking the wealth code is to understand whether there is something deeper, some behavioral wellspring of the 4DTPs that keeps a person on track.

Fortunately, Stanley did not leave the 4DTPs alone. He probed and he studied, and here we concentrate on two of Stanley's *hidden lessons,* lessons that help explain not only the key habits of successful people, but also how they kept at them. In particular, we focus on two hidden lessons revealed in Stanley's work, engagement and courage.

3.3   HIDDEN LESSON 1: ENGAGEMENT

The first hidden lesson is what we will call *engagement* or what Stanley called *vocation.* After careful analysis of the results of his study, Stanley found that many of the 733 millionaires worked in areas they found profoundly engaging, work they profoundly loved. Over and over again, the millionaires answered their questionnaires, interviews, and focus groups with tales of love for their work. Their work was not a job; it was a vocation or a calling.

Stanley's work in the literature of financial success is corroborated by work in the psychological literature of optimal experience. In particular, vocation may be mapped to Csikzentmihalyi's notion of *flow* (Csikzentmihalyi, 1990). Csikzentmihalyi studied high-performing individuals in a variety of occupations, from white collar to blue collar, artist to professional, scientist to humanist, and found commonality in the *intricacy* and *interwoven* nature of their activities. The common sports expression of being *in the zone* captures the feeling of flow for those who have played a sport and have felt that special feeling of doing no wrong. For our perspective here, the added idea of flow research is that optimal experience results from a complex of interwoven, engaging activities. Flow is engagement on steroids, but it is the accumulation of a large number of interrelated competencies that result in the sustained, high level of achievement.

The lessons of engagement and flow are rather interesting, and some might be tempted to call them yet other platitudes. Yet, love of work (engagement) and interwoven complexity (flow) are not particularly common themes. In fact, many people get it exactly backward. Many people say that they will work at jobs and tasks they don't like so they can earn enough money to do what they love; millionaires and successful, happy people the world over know this to be largely 180° out of phase.

To crack the code of engagement, and start toward a life full of flow, first we consider why engagement matters so much, and then we examine how to actively pursue engagement in our lives through the use of a simple, practical, yet reliable, test.

3.3.1   Why Engagement Matters

Once we acknowledge engagement as important, we need to question why it matters so much. Two explanations come easily to mind:

   1. Time flies when you're having fun.

   2. Details aren't annoying when they are part of an enjoyable bigger picture.

We've all had the experience of time "flying" when we were engaged. When was the last time it happened to you, and what were you doing? For me, my most recent time-flying experience is this current one: As I write these lines, the hours are passing, but I hardly notice. In other words, for me writing is engaging, and being a college professor is probably a good line of work for me.

How about you? Maybe time flies for you when you work with your hands, program a computer, read, write, do a tough problem, or participate in a favorite sport or hobby. It really doesn't matter which activities cause time to fly for you (within the bounds of legality and morality); it does matter that you know which ones they are and try to find ways to make them a part of your work life.

Of course, the knowledge that time does fly when you're having fun is no secret, but why is it a determinant of success? Time flying breeds a kind of unconscious patience that keeps you on track and makes the 4DTPs work for you. Whether you build a business or climb the corporate ladder, you must develop your personal and organizational competence before success will knock on your door—and that development takes time. Considering how impatient human beings are, it is that much easier to stick to the task at hand if the passage of time doesn't seem so tedious.

Another thing about being engaged is that it takes the devil out of the details. All jobs worth doing have myriad details that need to be done right—and guess what? People who are engaged with their work don't mind doing them because they are involved with getting the total job done and therefore have a good attitude toward what otherwise could be tedious labor. By way of contrast, the disengaged worker finds the details annoying and wishes the work could be delegated away. Although this lesson applies to all jobs and all workers, as individuals we can only exercise reliable control over ourselves (and even that is often in doubt). Therefore the best bet is for each of us to strive to find a work situation that offers the opportunity to make a contribution to an effort that engages us. Situations that match our inclinations aren't altogether easy to find, but they needn't be that difficult either.

3.3.2 Matching your Vocational Impedance

The lesson of engagement is both liberating and vexing. It is liberating to know that the road to wealth is paved with happy effort, but it is vexing to try to find our way to work that makes us happy. How can we better match our inclinations to the work we take up? This question is important in finding what we'll call our *vocational impedance match*. Just as electrical engineers know that systems work best when the impedance values of two interconnected subsystems match, in the real world we plug into life better if our interests and inclinations are consonant with the demands of our work. Here, we briefly examine a practical method to help inventory the activities you might want to consider in finding engaging work.

Time flies when you're having fun; but what activities keep you jolly, and what skills or tasks make up those activities? An entire aptitude-testing industry has arisen to answer such questions, but our approach to uncovering our inclinations will be much simpler. Let's write our way to an answer with the following exercise.

Exploration Exercise

Write an essay that explores three specific instances when time flew and three instances when time crept. Do not limit yourself to work activities. Include activities around the house and leisure activities as well. Explore the various skills and activities involved in each instance, and try to assess which contributed to time's flying or creeping.

The "time flying and creeping" criterion is the key to this exercise because it gets you to focus not on whether you were consciously happy but on whether you were engaged enough that you did not notice the passage of time. As a secondary criterion you can evaluate how annoying details seemed. These two guidelines should help you perform some meaningful introspection.

3.4   HIDDEN LESSON 2: COURAGE

Engagement with one's work was a key lesson behind the success of the millionaires in Stanley's study, but it wasn't the only one. Stanley found that millionaires are often courageous in moving forward in the face of criticism, risk, or other obstacles that might make others flinch.

On its face, courage isn't a particularly surprising attribute because many of the millionaires took risks to start businesses, walk away from successful jobs, persist in the face of failure, and persist when others told them they were being foolish. But the label "courage" doesn't really explain *why* some are able to ignore these difficulties and why others are not. Fortunately, psychologists have studied this topic under the rubric of locus of control.

3.4.1   Locus of Control: Internal versus External

Julian Rotter's work on *locus of control* (Rotter, 1954, 1966) is part of a larger *expectancy value theory* that argues that behavior is determined not by the size of reward but by a person's beliefs about the likely outcome or results of behavior. Although finer distinctions can be made, Rotter argued that two large categories of belief can be discerned. People can be classified as having internal locus of control or external locus of control depending upon whether they believe behavior is largely determined by their own actions or by the actions of others or some outside agency.

Various tests have been devised to measure whether a subject is an internal or an external, but the issue here is to map locus of control to courage and understand the connection. Ultimately, courage is a form of internal locus of control. When someone is courageous, they are largely betting that (a) they are right and (b) they can control events sufficiently well to succeed. If someone is constantly buffeted by outside agencies or the beliefs of others, they are unlikely to make a decision that goes against the received wisdom.

Seen in this way, internal locus of control is a precondition of courage, but perhaps the more interesting issue is whether courage can be learned. To turn the question around, we know that helplessness can be learned from Martin Seligman's pioneering work (Seligman, Maier, & Geer, 1968). While studying the relationship between fear and learning, Seligman discovered that dogs could be taught that to try to escape an electric shock was futile. Dogs were first restrained while receiving the sound of a tone directly followed by an electric shock. Next Seligman and several of his colleagues left the dog unrestrained and expected the dog would flee at the sound of the bell, but to their surprise the dog remained at the signal and took the shock. The results of this experiment, applied to humans, led Seligman to conclude that what one expects determines one's behavior. Seligman referred to this as *learned helplessness.* The dogs could be conditioned fairly quickly to believe their own actions mattered little and that their comfort was determined entirely by an outside agency. In Rotter's terms, Seligman discovered that dogs could be taught to become externals quite easily.

This is interesting, but we are less interested in helplessness and more interested in whether we can learn to take better charge of our lives, and Seligman and others have been concerned with these more positive aspects of human behavior for the last several decades. The take-away lesson of the *positive psychology* movement (Seligman, 1998) is that people can change their beliefs and explanations for the events that happen in their life. A growing body of research supports the notion that a primary determinant of our behavior and our happiness is our *explanatory style* and that those of us who habitually explain matters optimistically are happier and more successful than those who are habitually pessimistic. In a similar vein, it isn't much of a stretch to conclude that human behavior can be modified toward a form of *learned courageousness.* Although Stanley's data doesn't address whether his millionaires learned to be courageous or merely had a natural propensity toward it, given the mental flexibility of our species, it seems reasonable to assume that marginal increases in courage can be learned and may be important to increasing our chances for future success.

3.4.2 Exploring Courage

In general, do you believe you largely control your own destiny, or do you believe that your destiny is largely shaped by others? The following exercise asks you to explore these matters by writing a short essay.

Exploration Exercise

Write a short essay that explores two specific instances in your life, one when you went along with the crowd and the other when you went your own way. Consider the outcomes of both cases. Were the results favorable to you? Describe the ways in which the instances are or are not representative of the way you now generally handle such situations. Do you believe you have largely internal or external locus of control? If you could change this aspect of your personality, would you, how, and why?

3.5 TACTICAL LESSONS OF HANDLING MONEY

Engagement and courage may be thought of as key strategic dimensions of wealth creation, but what about the blocking and tackling of handling money? Here three issues draw our attention: How we spend money, how we earn it, and whether we save it.

3.5.1 Spending and Earning Styles

What kind of spender are you? Do you get a couple bucks in your pocket and feel the urge to go shopping, or do you save and save and then have difficulty buying something you really need? Most likely you're somewhere in the middle of these two poles. However, to simplify matters, in the language of Las Vegas, we'll call the first category of spender a *high roller* and the second category a *low roller*. Although "high roller" and "low roller" may have negative connotations, here the terms are used merely to be memorably descriptive of spending style. As it turns out, the Stanley study does have something to say about the connection between spending style and the type of wealth one might accumulate (balance sheet or income statement affluence), but for now, our purpose is to consider spending in the light of earning.

Having considered the spectrum of spending styles, can we similarly categorize the ways in which we earn money? Some thought suggests that there are basically two ways people typically earn money: in *clumps* and in *trickles*.

For example, commission salespeople get paid in clumps: Make a big sale, get a big paycheck. Freelance writers, artists, and consultants are paid in similar lump-sum fashion. The person starting a business has periods of clumplike payment, although after companies get underway, they themselves may be divided into clumpish and tricklish types of companies.

On the other hand, salaried men and women see their wages in trickles; every week, every 2 weeks, or every month they receive a modest but regular sum of money. These are two very different ways to receive money, and neither is intrinsically better than the other toward the creation of wealth.

3.5.2 Spending–Earning Impedance

On the one hand, people can spend as high or low rollers and they can earn in clumps or trickles, but just as we attempt to match the tasks we find engaging with the work we do, it makes psychological sense to seek a match of our spending and earning styles. That is, high rollers should be more comfortable with clumps and low rollers should be more at ease with the trickle. High-roller engineering graduates working as salaried design engineers and low-roller engineers who chase the big bucks in technical sales may be equally unhappy because both types ignore important personality facets in their career decision making.

Of course, there are hybrid earning styles combining both clumps and trickles, and these can be satisfying in providing both the consistency of the trickle with the upside potential of the occasional clump. For example, regular jobs that permit outside writing, speaking, and consulting offer this hybrid combination, as do regular positions combined with outside investing in real estate or financial markets.

Exploration Exercise

Write a short essay where you consider three recent purchases: one minor, one moderate, and one major. Consider the decision-making process that went into each purchase. Was it impulsive, considered, or agonized? How long did the decision take? Did you comparison shop? Did you buy top of the line or bargain basement? After considering the three purchases in some detail, classify yourself as a high or low roller.

Exploring vocational and earning–spending impedance are challenging matters, and I should warn that preferred skills and activities—and spending style—can change over time. Old dogs can learn new tricks; but even so, performing an analysis such as this, now and at intervals throughout a career, can be helpful in making effective adjustments.

3.5.3 Investing, Saving, and Thrift

For some reason that I've never understood, there's a lot of pressure to invest one's money in stocks, bonds, puts, calls, futures, bellies, straddles, or whatever the financial instrument *du jour* happens to be. Perhaps this pressure comes from a self-interested financial community, or perhaps it is natural in a capitalist economy—but, whatever the reason, the pressure is there. It is so ingrained that sometimes we are driven to the mistaken conclusion that investing is the primary way to wealth and that work is the way one sustains the habit of investment.

For many it is a costly habit indeed. As we've seen, wealth is often the result of sustained engagement when an individual is engrossed in his or her primary activity. Subsequently, the now-wealthy people have money to invest and do so. The problem is that the rest of us observe the latter phase—nobody paid much attention to these people until they became wealthy—and mistakenly think that their current behavior is responsible for their great wealth. Investing as a primary way to great wealth makes sense for engaged investors, but for the rest of us, investing is more of a sideshow than the main event. This is not to say that we should not take care of our money, and we turn to the data to see what it tells us about savings and thrift.

As a group, Stanley's millionaires were a thrifty lot. They drove modest cars, lived in modest homes, and were conservative spenders. They preferred the established over the trendy. They preferred to resole shoes, reupholster couches, and stick to weekly shopping lists at the grocery store. Further, while millionaires invested in the market, often doing their own market research and stock and mutual fund selection, they frequently invested in their own businesses, and many of their investment decisions were driven by a love for their vocation. Moreover, the lifestyle of the millionaire is not that of the jet-setter; it is the opposite. Topping the millionaires' list of nonwork activities is time spent with family and close friends, studying investments, exercise, attending lectures, and do-it-yourself home improvement projects. These spending and lifestyle activities fly in the face of widely held stereotypes, and although nobody is suggesting that these behaviors should be slavishly copied to increase one's chances of becoming wealthy, it is certainly a mistake to copy stereotypes of the wealthy when those stereotypes are not statistically correct.

Perhaps the lesson for us here is that everyone can benefit by developing the habits of *savings* and *thrift*. Before taking a flyer on Peruvian grape futures or other exotic speculations, it makes sense to accumulate an amount of money that equals or exceeds a significant proportion of your annual salary as a way to protect yourself against unexpected financial problems. Moreover, having cash in the bank gives you a cushion should you choose to switch jobs, go back to school, or otherwise do something that interrupts your income stream. In thinking back to my own decision to return to graduate school after having worked for a number of years, I know that it was made easier because I had stashed away a nice nest egg.

Once you've saved some money, it's fairly natural to ask whether you are getting a reasonable return. One of the best returns on investment comes not from an investment but from smart shopping. To understand this, consider that if you can get 25 percent off the price of a $100 pair of shoes you need, that savings is equal to the amount of interest you would have received had you invested the $100 in an instrument bearing 25 percent per annum. There are few (legal) investments that regularly pay 25 percent per year, but there are many opportunities to save 20 to 50 percent on items that we need every day. Of course, one of the tricks to this is not letting your thrifty consumerism lead to unnecessary or unplanned purchases; nevertheless, thrift in purchasing certainly can help put money in the bank.

To understand what to do with your money once it is in the bank, it is helpful to develop mental models of financial investments and how they work. To do so requires reading, and *The Wall Street Journal* (WSJ) is a good place to start. The WSJ is a good newspaper on its own merits, but if you become interested in things financial, it is *the* newspaper. *Forbes* is a good choice among business magazines because the writing is lively, the news is current, and the tone is irreverent. The classic book in economics self-education is Hazlitt's gem, *Economics in One Lesson* (Hazlitt, 1979) and Sowell's more recent volume (Sowell, 2004) provides a somewhat more elaborate view in plain English. To get an overview of investing, read Tobias's sensible little book, *The Only Investment Guide You'll Ever Need* (Tobias, 2005).

While investing money is secondary to developing core competence at engaging work, there are basic decisions to make in the allocation of money. A good approach is to read and engage a number of the suggested references or other current texts on these topics.

3.6 GET A LIFE

Engagement with your life's work is important, but I should hasten to say that this does not preclude a family life or other outside interests. All work and no play does both a stale engineer and a lousy family person make. Of course, there will be times in your life when work will be quite demanding and outside interests will suffer, but such times will be balanced by others when family duties call and work has to take a back seat. Knowing how to manage this shifting balance is difficult, and there is little that can be said in general about these matters, except that we should try to avoid using one of life's facets (such as work or family) as an excuse for neglecting another.

Even when we are able to live a fairly balanced work–home life, the demands of juggling many balls require that we manage time well, particularly at work; the suggestions of another chapter can be helpful here. Many workaholics—far from being time efficient—base their long workdays on a foundation of poor use of time. Knowing that you're going to be around later takes the pressure off doing something now. On the other hand, a balanced existence that forces you to divide the day among various facets of life can actually help you to better allocate your time in each facet.

One way to squeeze more out of your day is to rearrange your sleep schedule. For example, Ben Franklin rose at 5 a.m. (Franklin, 1791/2004); that same habit can be useful in more modern times. By rising at 5 a.m. and getting to work shortly thereafter you can (1) have uninterrupted time to write and think first thing in the morning, when you are freshest, (2) stay until a normal quitting time and squeeze more out of the working day, and (3) spend a full evening with your family or friends without worrying about work. Others rearrange their schedule for late-night activity, and there is some evidence of a physical basis for both larklike and owlish behavior, and decisions about scheduling are very personal matters. Nonetheless, sometimes our sleep habits are dictated by external factors or conventions rather than conscious thought; some thought about your own circumstances, followed by a conscious decision, might generate a schedule more suitable to your needs.

3.7 PLOTTING YOUR COURSE: VALUES, MISSION, AND GOALS

Television commercials and corporate annual reports are awash in "vision" and "mission" statements, and sometimes the main value of these public displays is to highlight unfavorably the difference between corporate word and deed. Nonetheless, at a personal level, shifting from a direct concern for wealth to one of engagement naturally leads us to think in terms of our own *values*. In other words, what set of concerns drives us in life? What categories of activity bring personal meaning and fulfillment. Articulating personal values is a useful exercise in self-knowledge, and once this is done, it is possible to write a *personal mission statement*. Just as companies write mission statements to inform stakeholders of their intentions and to drive decision makers to do the right things, personal mission statements can help guide individuals through the thicket of long-term temptations and opportunities to arrive at a more meaningful life. While a sense of mission helps us keep our eyes on the long term—mission is the strategic side of personal planning—*goals* keep us moving incrementally toward fulfilling our mission—goals are the tactical guideposts for near-term action. An understanding of the role of engagement naturally leads us to considering values, mission, and goal setting.

3.7.1 Creating a Personal Values Statement

Seeking engagement by paying attention to those activities that make time fly for you is a reasonably reliable means to finding categories of activity or behavior that we *value*. The subject of values can be a difficult one because of its ties to deeper issues in moral philosophy, but here we simply seek to raise the stakes on the detailed explorations that arise from time-flying analyses to a somewhat more general listing of personally prized categories of behavior to create a *values state- ment*. In other words, given your preferences and proclivities, what categories of activity and behavior do you prize most highly (Seligman, 2002)? Of course, we need to be a little bit careful and move somewhat into the realm of moral philosophy. If you were to say that you prize murder for hire and its efficient execution most highly, few readers (or courts) would condone or even tolerate your choice. So, more to the point, a personal values statement is a listing of *vir- tuous* (at least legal) activities or behaviors that you value most highly (Peterson & Seligman, 2004).

A good template for a personal values statement is to have a short preamble followed by an amplified list of values in *newsletter format*:

Values Statement for John Q. Doe

The following is a list of my core values in roughly decreasing priority:

Family Collectively, my family is the most important institution in my life, and individually Jill, Fred, Sam, and Roxanne are the most important people in my life.

Security The world is increasingly a dangerous place, and my family must have a secure home far from potential upheaval and unrest.

Learning I am a learning machine and am happy when I am engaged with unfamiliar or new ideas across a broad range of topics.

Order I seek order and try to leave things in better shape than when I first arrive.

Fitness I seek bodily health and fitness and bring balance to what would otherwise be a cerebral existence.

Financial security Although I am not driven to seek great wealth, I do not turn away from it, and I use money to advance other primary values to the benefit of my family, myself, and others.

Spirit I seek spiritual understanding and inner peace through reflection and religious practice.

Creating your first personal values statement is a difficult exercise that should probably span several days, weeks, or even months. In writing the values statement, it is important to list things you truly believe in. Sometimes people have a tendency to write platitudinous values statements that pay homage to things *everyone* is supposed to believe in. Motherhood and apple pie have their place, but the writing of a values statement is no place for self-delusion. After creating a draft, ask yourself point by point whether each item really matters to you. Putting it somewhat differently, would you sacrifice something important to live up to the value, or are you merely mouthing words so that others might think better of you?

Of course, values statements may be useful to try to change yourself, and you may write a value that you would like to emphasize in the future. Such items on your values list have a place, but be sure you are truly committed to doing what it takes to alter your behavior and embrace the value. Simply mouthing the value, placing it on your values statement, and pretending it is yours is not enough.

Once the list is created, it is useful to revisit it on a regular basis, perhaps annually. Where lists of objectives and specific goals will change frequently, your values probably won't change as rapidly over time. Of course, growing older has a way of changing a person's perspective, and it is common to add items to your list as different values surface or simply become apparent over time. Additionally, at key junctures in your life, you may find certain values ascendant and others somewhat less important to you. Keeping and updating a values statement throughout your life can help you recognize these junctures and understand the forces in your life that are shaping the changes.

Exploration Exercise

Write a personal values statement. Show the statement to a spouse or a very close friend and seek feedback on whether the values you have written reflect the person's knowledge of you. Revise the statement accordingly, taking into account the feedback you have received and how it accords with your knowledge of yourself.

3.7.2 Writing a Personal Mission Statement

The writing of corporate vision and mission statements has become so common place it has become something of a cliché. Yet, the impulse behind all the visionary and missionary ink is well intentioned. The best corporate mission statements succinctly reflect a corporation's values by articulating (1) the company's reason for existing and (2) the targets and aims of its activities. Many companies benefit from the writing and distribution of a mission statement and the reasons are mainly fourfold:

1. The thinking that creates a mission statement helps clarify what is truly important for future success.
2. Articulating mission creates a set of *ideals* to help guide future corporate conduct.
3. Articulating mission to employees and stakeholders announces clear targets to those whose performance is needed for success.
4. Announcing mission to customers and the public places pressure on the corporation to *close the gap* between corporate ideals and reality.

For an individual, a *personal mission statement* serves these same functions. Thinking clarifies values and targets, the announcement creates an ideal, telling yourself is a reminder of right conduct, and telling others encourages you to close the gap between the way you would like to be and the way you are.

A personal mission statement should be a *succinct, memorable* list of numbered points that are *broadly drawn*, yet reasonably *concrete*. The list should be preceded by a *preamble* that can be as simple as a single phrase ("My mission in life is to") or a more elaborate, yet brief discussion of key assumptions or background. Careful editing and word choice is necessary to embrace your mission succinctly, memorably, and accurately.

Mission Statement for John Q. Doe

My mission in life is threefold:

1. I must raise three citizens of the world who can thrive in an increasingly competitive, information-driven, and global economy.
2. I must love and understand my life partner as we face life's challenges and grow together.
3. I must create new products that have never existed using a balanced understanding of engineering knowledge, business skill, and markets and customers.

As with a values statement, mission statements should not say things because they sound good or make you seem more high minded. Instead make sure you believe in the mission points by asking yourself what activities in your life you might be willing to forego to live up to the particular point you are considering.

Getting feedback from a spouse or a close friend on your draft mission statement is useful, and posting the mission statement with that person helps create external pressure for you to live up to your intentions. As you review your values statement and goals, you should review your mission statement to make sure that it remains a valid snapshot of where you believe you need to go.

Exploration Exercise

Write a personal mission statement as a list of mission elements preceded by a preamble. Challenge each mission element. Is it a platitude or is it something you believe in strongly? Give the draft statement to a close friend, parent, or spouse for feedback. Carefully edit the final statement so that it is accurate, succinct, and memorable.

3.7.3 Setting Goals

At a strategic level of personal planning, value statements lead to mission statements, but to actualize the lofty aims of the mission statement, it is important to visit and revisit your *goals*. Goal setting feeds your to-do list with activities that are worth doing—with activities that help you accomplish your life's mission.

Goals are action-oriented statements describing desired behavior resulting in specific accomplishment. Goals should be *concrete* and *measurable*: They should describe the behavior or result desired not the process used to achieve it. Additionally, target dates should be given with each goal.

A good format for goal lists is newsletter format.

Goals for John Q. Doe January 11, 2006

Weight and fitness Lose 10 pounds of fat and gain 5 pounds of muscle. **6/2006**

Host/client project Finish host/client project for Mismara Corp. on time and under budget. **12/2006**

Education Educate children by seeking best schools, teachers, and extracurricular activities possible. **Ongoing**

Promotion Seek promotion to team leader through hard work, night classes, and pursuit of excellence in personal and team performance. **12/2007**

Jill Find activities to do together to keep marriage strong. Investigate ballroom dancing classes, co-ed yoga, and other active recreation. **3/2006 & ongoing**

Graduate degree Seek graduate degree combining technical and business-related subjects as credential for advancement and to build knowledge useful at work. **Start 9/2006–Complete 9/2009**

The newsletter format makes the goals easy to review, and the imperative verb form at the beginning of each sentence gives the goals a kind of force that is useful in helping a person get off the mark and really accomplish the goals.

Exploration Exercise

Write a list of personal goals using a time horizon of between 2 and 5 years. Review each goal and make sure each one is concrete, measurable, and time specific.

Goals should be reviewed on no less than an annual basis. Keeping a list of goals in a prominent place helps keep them in mind. Reviewing them each year helps you to see how successful you are in checking them off. Oftentimes there are good reasons why a person may need longer than anticipated to accomplish a goal, but if you make such lists and check them regularly, you will find that your life is better directed toward the things that you believe are important.

SUMMARY

The chapter started by examining four common-sensical attributes associated with the wealthy: (1) integrity, (2) discipline, (3) social skills, and (4) hard work. But common sense is one thing and common practice is another, and the chapter continued by considering two hidden lessons of the wealthy: their engagement and their courage. The wealthy are able to reliably exhibit integrity, discipline, social skill, and hard work because they often are fundamentally engaged with a life's work or vocation. Moreover, the wealthy are often courageous in the face of criticism and failure.

Learning to be more courageous is difficult, but by paying attention to activities that make time fly, all of us can better match our vocational impedance. By seeking career possibilities that match well with those activities we find engaging, we increase our chances of success.

After tackling the twin strategic issues of engagement and courage, the chapter turns to tactical matters of handling money: earning, spending, and saving. It also considered the importance of balance of work and home life.

Having flipped the money chase on its head and emphasized engagement, the chapter has gone a step further and considered the importance of exploring who you are through the systematic creation and updating of values statements, missions statements, and lists of goals throughout your life. These exercises can help beyond the day-to-day bustle of work and chart a course for career and life success on terms that are personally meaningful. Exploring and understanding these matters is not easy, but reasonably modest expenditures of time and effort in these activities can have a substantial payoff over the course of a career and a life.

EXERCISES

1. Interview a successful person. Ask that person questions about his or her early years and about how he or she became successful. Ask questions about that person's present work life. Write an essay discussing the keys to his or her success. Compare and contrast levels of engagement and enthusiasm early and late in the person's career.

2. In a short essay, consider whether your current career or career plans help create a personal impedance match to your inclinations and spending style. If so, which elements of your goals should be emphasized to enhance the impedance match? If not, what career options might offer a better impedance match and why?

3. Form a team of two. Each of you should write a brief statement about time-flying activities and interests. Exchange lists, and then write a brief essay on specific career options that seem to match the other person's list. Read and discuss both sets of recommendations.

4. Form a team of two. Each of you should make a list of major purchases made over the last 2 years. Exchange lists and write a brief essay classifying the other person as a high or low roller. Discuss the essays.

5. Form a team of three or more and choose one individual whose career choices will be examined during a brainstorming session. Try to find alternative career paths or emphases that will help the individual to grow. This exercise can be repeated for each group member.

6. Form a team of five and have each person make a list of major purchases made over the last 2 years. Make copies of all five lists and circulate them among the group members. Rank the lists from highest to lowest rollers and discuss attitudes toward spending money.

7. Make a list of personal values and prioritize them.

8. Write a personal mission statement.

9. Make a list of personal goals over the next 5 years. Be specific. Project dates of accomplishment and, where possible, quantify the goal.

10. Consider whether you have ever known someone who tried to socialize his or her way to wealth. List the ways in which they were successful and whether that success appeared sustainable? List a specific example or examples when their efforts failed.

11. Consider whether you have ever known someone who tried to invest his or her way to wealth. List the ways in which they were successful and whether that success appeared sustainable? List a specific example or examples when their efforts failed.

Chapter 4

Getting Organized and Finding Time

4.1 TIME AND ITS LACK

Time is one resource we never seem to have enough of; when it runs out, there is no more to be had. Yet, despite its importance and scarcity, it is remarkably easy to waste. To *Homo sapiens*, procrastination is as easy as eating, sleeping, and breathing; even for the engaged entrepreneurial engineer, getting the most out of the working day is a difficult challenge. To make matters worse, it is remarkably easy to blame our time wastage on external factors and overlook the enemy within. Our clients, our bosses, our families, and our friends are easy scapegoats; although they and others are sources of added work and interruption, the real villains are our own lack of organization, our own lack of discipline, and our own misunderstanding of the pivotal role time plays in our lives. To combat these difficulties, we must examine the many ways people waste time and consider key techniques for gaining control of our schedules and of our lives.

4.2 EFFECTIVE WAYS TO WASTE TIME

The wasting of time is an old and venerable activity. Long before the beginning of recorded history, our cave-dwelling ancestors spent time rooting around the cave trying to find the mislaid flint stone so they could start the fire. After many a cave meal many a cave spouse had to nag the procrastinating other spouse to take the carcass outside. With civilization came new technological achievements and advancements in social structure that have helped raise time wasting to its current high art. With alphabets came the opportunity for junk reading, and with movable type came the opportunity for high-volume junk printing. The establishment of regular postal service opened the door for delivery of that junk printing as junk mail. The telephone and the computer have opened new vistas; we now have untold opportunities to send megabytes of useless trash around the world at the speed of light.

The Entrepreneurial Engineer, by David E. Goldberg
Copyright © 2006 John Wiley & Sons, Inc.

Social organization has been no less successful in improving the opportunities to waste time. Kin groups led to tribal organizations to nation-states, to corporations, and, finally, to that most time-waste prone of all organizations, the university.

Over these tens of thousands of years the variety of ways to dispose of every spare moment has grown tremendously; nevertheless, it is possible to categorize fairly broadly the ways to waste time:

1. Misplacing things

2. Procrastinating

3. Task switching

4. Never saying no

5. Reading everything

6. Doing everything yourself

In the remainder of this section, we examine these techniques and their effectiveness in some detail.

In a business that generates more than 10 pieces of paper a year, one of the most effective ways to waste time is *misplacing* documents you'll later need. Some individuals are quite systematic in their efforts to misplace important documents. These pile drivers have developed an especially effective means of losing any document through the utilization of the pile document retrieval system (PDRS). In this system, the user creates several 3-foot-high piles of recent and not-so-recent documents. When faced with a need for a particular document or approached with an information-retrieval query ("Do you have memo X?"), the PDRS adherent wheels around to the piles and utters four magic words: "It's in here somewhere." Ten minutes later the PDRSer promises to send a copy when he or she finds it. Sure—and the check is in the mail.

Another proven means to waste time is flat-out *procrastinating*. What the art of procrastination lacks in subtlety it makes up for in unrelenting ability to avoid even the simplest chore. There is room for difference of opinion on this matter, but I feel some of the most creative procrastinators today are those who practice the art in the name of "time management." These people adopt time management schemes with impossible prioritization plans, using multicolored pens, fancy notebooks, or the latest in personal digital assistant technology, only to tell you why the simplest task can't be accomplished in under two fortnights. It is difficult to imagine that procrastination might become any more refined than this, but we shouldn't underestimate the innovative capability of our species.

An equally useful yet somewhat more subtle time-wasting technique is that of *task switching*. Because most jobs require some time overhead to start or restart, task switching maximizes time spent on overhead activities and minimizes time spent on productive ones. When combined with a telephone ringing off the hook and co-workers wanting to talk about the White Sox or the Bears, task switching can achieve near-zero rates of productivity. At the same time, it is the rare task-switching pro who can't get sympathy from co-workers and friends by complaining about all of the balls he or she must juggle.

Another way to make sure you rarely accomplish anything is *never saying "no"* when you are asked to do something. Even modest-size organizations have a large number of people sitting around with nothing better to do than to generate forms, surveys, report requests, and other trivia to occupy one's time. A perfectly reasonable time-killing strategy is to take all these requests seriously. Fill out that survey on company recreational policy; answer that letter regarding a charitable contribution to Poodles Without Puppies. The skillful practitioner could spend an entire career on information exchanges no more urgent than these.

A close relative of never saying no is *reading everything* that crosses your desk. Important documents, like that 10,000-word article on Zimbabwean frazil ice, should not only receive a close reading from you but may even require a detailed proofing. Not only will this activity dispose of unneeded time, you'll have great fun in recalling third-grade glory days when you were spelling-bee runner-up.

The classic way to prove that you've arrived as a time waster is to try your hand at *broom grabbing*. This maneuver requires that you first hire good people and then do their jobs in addition to your own. This ploy earns bonus points for the successful stylist because you not only waste your entire day trying to do the work of your subordinates, you completely demoralize and alienate them in the process.

Though my tongue has been firmly in my cheek for much of this section, the satire belies a sympathy for our human condition and our propensity to waste time. We have all been pile drivers and procrastinators, had trouble saying no, read the unworthy, and grabbed the broom. The simple truth is that these time-wasting habits are some of the easiest counterproductive habits to acquire and among the most difficult to shake. Since our natural inclinations work against us, we need a helping hand, a guide to self-discipline; in short, we need a system. A comprehensive, seven-part system can help us gain control of our schedules. Although we have no control over the passage of time, we can control its use.

4.3 SEVEN KEYS TO TIME MANAGEMENT

With so many opportunities to waste time, it takes a special kind of systematized vigilance to gain control of our schedules and become as productive as possible. In this section, we consider a seven-point plan of attack against the forces of time wasting. The plan involves a two-pronged assault, a pincer movement combining rear action against physical disorganization and a frontal attack against the enemies of productive time use. In this way, we can hope to fight off loss of time and become as productive as we can.

Specifically, the seven parts of our fight against time wastage are these:

1. Find a place for everything and have everything in its place.
2. Work for Mr. To Do.
3. Sam knows: Just do it.

4. A trash can is a person's best friend.

5. Tune your reading.

6. Manage your interruptions before they manage you.

7. Get some help.

When these elements are used together, they are powerful medicine against the disease of time misuse.

4.3.1 A Place for Everything

The electronic office is upon us, and far from earlier promises of the elimination of paper, we are awash in a sea of computer-generated reports and are snowed under a veritable avalanche of laser-printed mail and memos. Much of this stuf does not deserve a second glance (and I wish the first were somehow avoidable), but some of it is germane and needs to find a home. The easiest thing to do with all this stuff is to put it in a pile on your desk. As more papers come in they, too, get added to the pile. For a time such a pile-oriented filing system works because information retrieval from a short stack is not too involved; but as the stack grows, the search time grows as well. It doesn't seem like such a big deal if viewed search by search, but suppose your stack grows to the point where a search for a single document takes an average of 3 minutes, and also suppose (conservatively) that you need an average of 10 documents per day. That means that you spend an average of at least 30 minutes per day searching through your piles. Assuming 5 days a week and 50 work weeks a year, this translates into roughly 125 hours, or almost 16 working days a year lost to shuffling through your pile. Almost all of that lost time is avoidable if you build your time management strategy on the bedrock of a good filing system.

This draws us to an important conclusion: To use time wisely, create and use a filing system. The old proverb puts this in more memorable terms:

Have a place for everything and have everything in its place.

There are two reasons—one physical and one psychological—why this is such an important tactic for good time management. Knowing where things are virtually eliminates the pile driver's pile shuffle, making you that much more efficient almost immediately. The second reason is that by putting things in their place you eliminate the stress of being literally surrounded by pending work. Of course, you must make sure you have a good way of knowing what still needs to be done after work has been filed away—and we will examine one approach to that problem in a moment—but the act of eliminating clutter in your work space can help reduce your worry about the many tasks you need to get done and let you concentrate on the job before you.

I suppose there may be complex theories of how best to create a personal filing system, but the most important things are that you create one and that you use it. Whatever the system, it should be (1) organized in categories that match your work needs and (2) designed so it is easy to add new files. Over time, it may become necessary to reorganize and recategorize; this will become evident as categories become overstuffed or go underused. Our discussion here is oriented toward the storage of physical pieces of paper, but the same principles apply to the organization of electronic files, as well. Electronic clutter can be every bit as nerve wracking and time consuming as physical clutter, and the efficient entrepreneurial engineer does what he or she can to manage it.

As a concrete example, I have listed the major categories of my own personal filing system:

- Correspondence (by name)
- Student files (by name)
- Course files (by class and year)
- Short courses (by title)
- Projects (by title)
- Proposals; requests for proposals (by title)
- Papers (by title)
- Paper reprints (by title)
- Personal business (by topic)
- Departmental business (by topic)
- College business (by topic)
- University business (by topic)
- Papers by others (by serial number)

You may wonder why I've shared this in all its gory detail. I remember that when I got my filing cabinet, shortly after taking my first job out of school at Stoner Associates, I was curious how other people organized their stuff. How is correspondence filed? How are project and proposal filing done? The system I have presented here is a cross between the things I learned at Stoner and some things I learned from my dissertation advisor. None of it is profound, but sometimes it is easier to design the mundane from other than a blank sheet of paper.

Although the list has been tailored to the needs of a college professor, there are a number of categories of general use. For example, everyone receives correspondence, and a good way to handle it is to have a single category, Correspondence. This category has individual files for frequent correspondents (by correspondent name) and miscellaneous correspondence files for ranges of the alphabet (A–F, G–M, etc.), where an individual letter is filed by correspondent name in the file folder with the appropriate range of the alphabet.

Student, course, and short-course files are peculiar to my line of work, but project and proposal files are probably necessary in yours as well. It is often useful to distinguish between active and inactive projects and proposals, relegating the inactive kind to deeper storage.

Professors are expected to publish or perish (I prefer the more positive exhortation, publish and flourish), and I keep my original papers in one category and a fresh stack of reprints (ready to go out at a moment's notice) in another. Perhaps in your business it is important to have company literature ready to go out or perhaps copies of past reports or designs. Whatever is important to have available should probably be filed in its own category.

Often there are personal matters that require attention at work (salary review papers, benefits, etc.) as well as departmental and other organizational matters. I keep these in separate categories, and you may find a similar arrangement useful.

Finally, I should mention a word about filing technical papers. It is tempting to keep a file of papers by subject, and this is satisfactory for collections of 100 to 200 papers, but beyond that a more systematic method is necessary. Ben Wylie, my dissertation advisor, filed his papers by unique serial number and uses an ingenious system of cross-referenced index cards for retrieving papers by author, title, or subject. I've adopted the same serial-number system, but I use a computer database to cross reference the file records. I use that database online to help locate a needed reference at a moment's notice.

However you construct it, a filing system gives you a place to keep things out of your hair and a simple way to retrieve them. Take the occasion of the next exercise to plan a new or more appropriate file system.

Exploration Exercise

Plan a filing system appropriate to your current or future line of work. If you already use a filing system, make a list of its major filing categories. Consider what changes to your system would make it more useful to your current work situation.

4.3.2 Work for Mr. To Do

One of the worries you might have as a born-again time organizer is that if you file something in your spiffy new filing system—if you follow the put everything-in-its-place stricture—you'll be subject to another law of human behavior: Out of sight, out of mind. Without a systematic means of task tracking, this is a risk. The trick is not to depend on "mind" at all. The trick is to work for Mr. To Do.

We've all used to-do lists at one time or another, but the veteran time manager uses one with considerable zeal. Some time management books suggest elaborate prioritization schemes, multiple lists, fancy calendars, and so on. The tools of our trade will be much simpler: a monthly calendar, an $8\frac{1}{2} \times 11$-inch lined pad of paper, and a diary or their electronic equivalents. The calendar is simply used to record all events, appointments, and fixed-date deadlines. The pad of paper is used to track all activities, both to do and pending, as well as each day's activities. The diary is used to track who you talked to and what you did. Of course, in an age of ubiquitous personal computers and personal digital assistants, calendars, to-do lists, and diaries all have electronic counterparts. I prefer to do my calendar electronically and my to-do list and diary in pen and paper. Perhaps you find it more convenient to do them all electronically, all on paper, or in some hybrid of the two. Regardless, the important thing is to find an arrangement that is convenient for you and well used.

For my to-do list, rather than elaborately classifying or prioritizing, I list two types of activities: to do and pending. To-do activities are those I need to do within the fairly immediate future. Pending activities are those I would like to do or those that must be done sometime down the road. With this scheme, every morning I review my activities of the previous day and create a list of the current day's activities. As each activity is accomplished, I take great pleasure in crossing it off both the daily and the pending lists; at the beginning of each week, I make a new sheet, updating the to-do and pending entries.

The calendar keeps track of dated activities and deadlines. Whenever an activity with a definite deadline comes in, it goes on the calendar and on the pending list. I check the calendar at the beginning of each week to see which activities are coming due. These are placed on the to-do list for the week, keeping me up to date.

I use the diary to keep track of who I talked to and what I did. To make it simple, I try to make entries in the diary as they happen. Sometimes this results in a somewhat messy diary, but I don't waste time writing and rewriting the same information. I also let my diary double as a technical notebook and do calculations and sketches for new technical ideas there as well. I find that bound notebooks of graph paper with numbered pages work well for both technical and written material. Keeping a written and dated record of your technical ideas is also useful in a patent filing. Pages from a business-technical diary can be notarized, and this is helpful if there is a legal question as to when an idea was developed.

The benefits of this system are several. Compared to many schemes, it requires little or no time overhead. Keeping track of your tasks on a to-do list helps clear your mind of the clutter of the many things you have to do. As the use of the filing system unclutters your physical space, task tracking unclutters your mental space. It also gives you a psychological boost every time you cross off a job that's done. Perhaps more importantly, by getting you to face what you do from day to day, it allows you to get a better sense of your productivity potential and to be better able to choose those things that are really important to your work.

Exploration Exercise

If you currently use a to-do list on a regular basis, stop using it for 3 days. If you do not currently use a to-do list on a daily basis, adopt the scheme just described for a 3-day trial period. After the experiment, write a short essay comparing and contrasting your experiences with and without the to-do list. Include a discussion of your perception of productivity, and cite any physical evidence of differences you experienced.

I don't want to bias your thinking, but if I get lazy and leave my list for a day or two, I can hardly wait to get back to it and get control of my working life. One of the easiest ways to keep your list under control in the first place is to get the little stuff done and off the list as soon as possible, and that's the next topic.

4.3.3   Sam Knows: Just Do It

My 6 years at the University of Alabama were happy and productive, and there I had the privilege of working with a great group of people. One of the lessons I learned came from a memorable mechanics professor and retired Army Reserve colonel, Dr. Sam Gambrell. Sam had the cleanest in-basket of any person I have ever known. Sam's in-basket was so clean that a piece of paper didn't even think about hitting his basket before he had a response completed and shipped out. I must confess that, at first, I thought it was a little silly to be so ruthless about incoming small stuff, but over the years I consciously tried to be a little bit more like Sam, and I began to see the wisdom of his ways.

Now I try to handle things as they hit my in-basket. If they're little, I just do them; if they're big, I file them and list them with Mr. To Do. There is some judgment required here, but I find when I refuse to let regular little stuff pile up, I keep a clearer calendar for the big things that need my attention. I should also confess that I can't quite live up to the example set by Dr. Sam, and from time to time I do get backed up with little things I should have handled quickly. Nonetheless, I know what I should do, and I try to keep clutter out of my basket and keep work flowing out the door.

4.3.4   A Trash Can Is a Person's Best Friend

One constant in business life is that more tasks come across your desk than you can or should do. The easiest way to handle some of them is simply to refuse to do them. The Almighty made junk mail and memos, but he also made pitching arms and wastebaskets, and we should use them. When a potential task first crosses your desk, ask whether you really need to do it; of course, as the new kid on the block, being too fussy can earn you the reputation of being uncooperative, so the new guy or gal needs to tread somewhat warily here. Even so, pay little mind to the junk mail, phone solicitations, and cold calls from salespeople—unless the products they are offering are necessary to what you're doing. Otherwise, let that useful two-letter word—no–together with the wastebasket (alias "the round file" or "file 13") unclog your schedule as fast as they can.

4.3.5   Tuning Your Reading

To make the "go, no-go" decision to keep something or pitch it, we often have to read some document, brochure, or other piece of written material to get enough information to know whether something is important or not. In addition, many tasks in the course of the business day have a necessary reading component. On the other hand, just because some amount of reading is necessary does not mean that every business reading task requires the full attention you might put into the reading of a textbook or a novel.

In fact, business reading requires that you have different speeds, that you tune your reading to the task at hand. Unfortunately, years of engineering schoolwork have taught many engineers to read at a methodical and often fairly slow speed, to absorb, for instance, all the material in some fluids or electromagnetics text. In business, this approach is wasteful, and just recognizing that different materials deserve different levels of attention can help.

A simple way to tune your reading is to think of reading at three basic speeds:

1. Skimming
2. Scanning
3. Reading

When you skim, your eyes should move from titles, to headings and sub-headings, to figures, charts, and tables, perhaps taking in important introductory and summary sentences at the beginnings and endings of appropriate paragraphs. Before reading any document in more detail, use a preliminary skim to derive a road map upon which more detail can be charted during a second pass; this should become a regular habit with everything you read for business.

A scan is more comprehensive than a skim; it should cover all elements of a document, not word for word, but as fast as you can while still feeling that you have passed through the whole thing. Speed-reading books teach scanning techniques—for example, the S-curve wiggle down the page and sighting whole word groups instead of single words—and such books and courses can be useful in building your scanning ability. I disagree with some of the speed-reading literature when it makes the overzealous claim that everything can and should be read at scanning speeds. Highly technical material, legal documents, contract specifications, and the like must be read and reread at no faster than near-spoken speeds; and, to be honest, recreational reading is more fun at slower speeds. Nonetheless, scanning is effective for much business material.

I will use the term reading to denote your normal textbook reading speed. How fast do you read when it is important for you to remember all of the material that you cover? I think that speed-reading books are somewhat deceptive in this matter by implying that there is little loss in comprehension with faster scanning rates. They "cheat" by defining comprehension as the percentage of correct answers to a superficial multiple-choice quiz given immediately following a reading. Full comprehension cannot be measured by multiple-choice exams; it comes from deeper readings than are possible with a scan. The speed readers do have a point, however, and much of what comes across an engineer's desk deserves little more than the type of scan taught in their books.

Experiment with skimming, scanning, and reading in the following exploration exercise.

Exploration Exercise

Take a five-page document from work, school, or home and give it successively (1) a skim, (2) a scan, and (3) a reading. Time each activity, and after each write down everything you can recall. At the end of all three activities, write a short paragraph comparing the amount learned in the successively more comprehensive readings. Consider under what circumstances the incremental knowledge gained would be worth the additional expenditure of time.

In general, it is best to approach a reading task with a skim followed by a scan and, if the information is sufficiently important, one or more readings.

4.3.6 Managing Interruptions

Almost every time management book that I have read exhorts you to manage interruptions; in a vacuum, this is good advice. The time waste of interruptions is both direct—coming from the expenditure of time on things you hadn't planned—and indirect—coming from the waste of unintended task switching and the associated overhead required to get back to the task at hand. Therefore, all other things being equal, interruptions should be avoided. But all other things are never equal. You don't live in a vacuum; you live in a world of co-workers, clients, bosses, and family members, all of whom have some legitimate claims to a portion of your time. You risk being seen as unfriendly, uncooperative, or worse if you are overly zealous in the protection of your time. If your engineering degree is newly minted, these warnings are especially important because you are low person on the totem pole, and part of the reason you have been hired is to make life easier for your senior co-workers; being unavailable can be a step toward being unemployed.

On the other hand, disruptions can be managed—albeit carefully. There are a number of ways to hold them at bay. The phone is a primary interrupter, and forwarding your calls to a receptionist or secretary during key work periods can help control phone interruptions, at least for a time. In our electronic age, many people use an answering machine, although some are old-fashioned enough that they prefer that their callers talk to a real person, not a whirling strip of magnetic media. Either way, once messages are taken, you are in control and can decide whether and when to return calls. Of course, your callers may likewise be unavailable when you return their call, thereby starting a nice game of telephone tag. Electronic mail can be useful in this regard, allowing a message to get through to an individual without directly interrupting that person. On the other hand, e-mail opens all kinds of avenues for wasting time, and many electronic messages deserve the electronic equivalent of the round file that receives so much of your other kind of mail.

Unwanted visitors can be partially controlled by closing your door or putting up a signal or sign at your cubicle at times when concentration is essential. Here again some caution should be exercised because always shutting out visitors can cut you off from your co-workers. Another way to control unwanted interruptions is to find a hiding place where you can work undisturbed. Much of this book was written in coffee shops and libraries, away from the phone and from my desk. Again, such methods should be used sparingly, lest you gain an unwanted reputation for being unfriendly (or for frequenting coffee shops during business hours).

Within reason, then, interruptions can be at least partially managed, but it is important to monitor the effect of your efforts on those around you. If you control interruptions by disconnecting from your network, you've lost more than you've gained. On the other hand, if you can keep interruptions under control, you're going to accomplish more in less time and be a happier camper, having greater productivity and better control of your schedule.

4.3.7 Getting Help

Another stock piece of advice in the time management literature is to delegate your work to others. It is true that managers must give jobs to others, let them do the work, and avoid grabbing the broom. However, a new engineer is not going to have anyone to delegate anything to, and sitting around complaining about that fact is counterproductive. In fact, griping about lack of assistance is one of the worst ways to waste time. Getting the job done is job one, and to get it done requires learning the secrets of personal productivity.

Nevertheless, even the new kid may have some opportunities to save time by working through others. Secretaries can be helpful, but the old-fashioned secretary has become an endangered species. Moreover, in this day and age, with a computer on every desk, it's questionable whether the usual back and forth with a secretary on something like straight typing is the fastest way to go. I find it useful to work with well-trained people who know my system and preferences, but using pool staff can be less productive than doing it myself. Of course, whether this is true for you depends on how fast you type; just make sure that when you get "help" it really does.

Doing rough documents through dictation is another way to use the help of others to make your life easier. At first, using dictation equipment takes some getting used to, but after some practice short letters can be completed on the first trial and longer documents will require only modest corrections. Dictation can be especially useful in initiating a rough draft. The raw text generated by dictation can be edited, cut and pasted, and interpolated with new text in the generation of a first draft. As modern speech recognition has improved, talking drafts straight into a computer has become more prevalent and can open up opportunities for productive use of driving or travel time.

SUMMARY

In this chapter, we've examined ways to waste and to save time. The primary key to personal organization is to "have a place for everything and have everything in its place," which requires the establishment and use of a personal filing system where all your business papers can find their final resting place.

Once you have a place for everything, the next most important key is to work for Mr. To Do. Keeping a simple to-do list of current and pending activities, together with things on today's agenda, will force you to face your time use (or abuse) squarely; the list will act as a kind of higher authority to keep you accountable. Beyond these two key activities, a number of other techniques can be adopted to keep your to-do list more manageable and to prevent unnecessary interruptions from diverting you too much. Some caution is required in adopting time management techniques that affect others to make sure the activity doesn't weaken your relationship with the people important in your work life; if reasonable care is taken to monitor this, you can save time and be viewed as a team player simultaneously.

EXERCISES

1. Interview an individual whom you judge to be an effective time manager. Write a brief essay summarizing the key techniques used by that person to help make him or her more effective.

2. Interview an individual whom you judge to be an ineffective time manager. Write a brief essay identifying the key ways he or she wastes time.

3. Identify the elements of your process of personal organization. What elements do you use? Diary, calendar, to-do list, filing system, or other? How frequently do you use them? Write brief paragraphs identifying three ways in which your current organization encourages effective time use. Write brief paragraphs identifying ways in which your current system encourages time misuse.

4. Based on your answer to number 3, design a new process of personal management. Write a short report detailing the design and its rationale.

5. Based on your answer to number 4, implement the new process of personal management and use it for 2 weeks. Write a short report on its benefits and shortcomings. Based on that analysis, modify the design and implement the changes.

6. Write a paragraph identifying your stand on the potential conflict between time management and human relations alluded to in this chapter. Is there a conflict? Which is more important? It might be helpful to consider specific incidents, hypothetical or empirical.

7. For 1 week practice what Sam Gambrell preaches, doing little things immediately and filing others for further consideration. Write a brief paragraph describing your experiences.

Chapter 5

Write for Your Life

5.1 ENGINEERS, ROOT CANAL, AND WRITING

The average engineer would rather go to the dentist and have root canal surgery than write a technical report or a memo. This is unfortunate, as a large part of an entrepreneurial engineer's professional life is spent in writing technical communiqués of one sort or another. Although the widespread aversion to writing has a variety of causes, a large part of the problem is simply not understanding the process and elements of good technical writing. Many engineers receive little or no formal training in the kinds of business-technical writing they must use in their work lives.

To partially remedy this situation requires the mastery of a number of the basics of technical writing:

1. The prime directive of all writing
2. The primary exercise for developing good writing habits
3. The primary structure of all technical writing
4. The technical writer's best friends
5. Summaries and conclusions: knowing the difference
6. Titles and headings made easy

In addition, for many people writing is a less-than-beloved activity because it is so intimately tied to criticism, and it is important to understand this to unblock bad writing habits and become a more fluid, natural writer.

5.2 WHY MANY ENGINEERS DON'T LIKE TO WRITE

It's odd that so many thoughtful people dislike writing. After all, it is one of the few school activities that follow us from grade to grade and class to class, from the first year of elementary school to the senior year in college. How is it that such an important activity, one we have practiced and practiced, becomes so frequently blocked during our professional lives and is therefore so often dreaded or avoided?

The Entrepreneurial Engineer, by David E. Goldberg
Copyright © 2006 John Wiley & Sons, Inc.

Although there are probably many reasons why people don't like to write, one answer for many engineers may be summed up in a single word: *criticism*. From our earliest school days, writing is an activity associated with making mistakes and receiving subsequent criticism from our teachers. Whenever I think of my second-grade teacher, Mrs. Brown, I can still hear her scold: "That's not a sentence," "Don't use a comma there," and "Watch your spelling." Because writing is such a complex activity, requiring the coordination of so many technical and creative skills, it does require a lot of feedback—much of it negative—for the student to master its basic technical aspects: grammar, spelling, punctuation, and sentence structure. Unfortunately, many of the people providing this feedback are less than supportive in correcting mistakes; worse, they are often unaware that constant criticism can damage the creative impulse. As we grow, we like to think that we have been able to cast off the yoke of such criticism, the emotional sting of these minor traumas, but more often than not they chase us into adulthood.

In the case of writing, the ghosts of the Mesdames Brown of the world haunt us as we sit, pen poised over paper, immobile, not able to dash off so much as a phrase. And if we do dare to write, we are scared witless: We are afraid of making mistakes. How many times have you sat with a pen in your hand and ideas in your head, staring at a piece of blank paper? Then, when a sentence does finally pop out and onto paper, how often have you thought, "No, that's not right," and scratched the sentence out? Of course, this hasn't happened to you just once. Over and over, sentences pop out only to be cut down by Mrs. Brown's goblins: "That spelling's incorrect" and "Your structure's too simplistic."

With a head full of evil spirits such as these, how can a person write? It's a lose–lose proposition if ever there was one: You're frustrated when you don't write because the job needs doing, but you're frustrated when you do write because the words won't come out just right. As a result, it is quite natural to avoid—or put off—an activity that prompts such a mixture of conflicted, largely negative thoughts; under such circumstances, there is little hope of ever enjoying the writing process.

5.3   PRIME DIRECTIVE OF WRITING: JUST WRITE

Having identified the fear of criticism that blocks many writers, we are in a position to consider a straightforward remedy. The prime directive of writing is—drum roll please—to just *write*. This sounds obvious enough, but when I watch students sitting in class, I see almost as much crossing out as writing. As a result, the writing process becomes a tedious herky-jerky affair marked by fits and starts, but not much completed writing. Thus, the key to exorcising the ghosts and goblins of criticism is to *separate writing from revision*.

The "write, cross out, write, cross out" mode of writing is the result of trying to do too much at once. We try to revise our words at the same time that we create them. These activities, revision and creation, are distinct; some psychologists would even argue that they are processed in different hemispheres of the brain. Whether or not that is so isn't important here, but keeping separate these two very different activities is.

Separating writing from revision is a sensible suggestion—so sensible that I wish I could take credit for thinking of it, but I learned this valuable lesson from a book called *Writing with Power* by Peter Elbow (1998). At the time I read that book, I was trying to confront the largest writing assignment I had ever encountered—my Ph.D. dissertation—and I was dreading the very thought of approaching something so long and complicated. I was browsing bookstores, looking for hints, clues, anything that would help me become a more confident and productive writer. From the first reading, Elbow's diagnosis and prescription struck me as just right. Of course, *saying* we should separate writing from revision and actually *doing* it are two different things. One of the best techniques for developing good writing habits that can last a lifetime is a useful exercise called freewriting.

5.3.1 Freewriting

To eliminate the damaging legacy of criticism we must first and foremost learn to get our thoughts—whatever they may be—down on a piece of paper. Peter Elbow (1998) recommends a straightforward exercise for learning to do that. He calls it *freewriting*; as the name implies, it is writing that is not directed at a particular subject, project, or piece. Rather, it is simply an attempt to transfer thoughts to paper in an unfettered manner. To nuts-and-bolts engineers, the idea of writing without clear purpose may seem a little too "touchy-feely" to be of practical use. In a moment, we'll explore the rationale behind the exercise, but for right now we'll concentrate on its mechanics.

Freewriting is easy to do, the only physical requirements being a pad of lined paper and a pencil or pen. Thereafter, the process may be defined by a set of six rules:

1. Write for a predetermined, fixed period of time.
2. Do not stop and, insofar as it is possible, do not lift pen from paper until the time expires.
3. Do not cross out any writing, not one word.
4. Do not worry about spelling, grammar, punctuation, or structure.
5. Do not fix your mind on a particular topic, although if you remain on one topic that's all right.
6. If you become stuck, keep writing by repeating the same sentence over and over or by writing about how it feels to be stuck.

The rules are straightforward, and the best way to illustrate them is for me to stop talking and for you to start freewriting. So that you will be as uninhibited as possible, please realize that what you write during freewriting will be for your eyes only. You will not be asked to turn it in, nor will you be asked to share what you have written with others. In short, this is an exercise for you to learn about you. As a final reminder, don't stop, don't cross out, don't worry about grammar or usage, don't worry about topic, and if you get stuck, just write *something* over and over until you get on track again.

Exploration Exercise

Freewrite for a timed session of 6 minutes. Follow the six rules of freewriting zealously.

How did that feel? I can recall my first encounter with freewriting. What a strange experience it was! I remember having trouble restraining myself from crossing out, but soon the words began to flow. I also recall being surprised by some of the thoughts that started to surface, but most of all I remember the liberating feeling of being "allowed" to let different associated ideas flow one from the other. I'm sure your experience with, and feelings about, freewriting will differ, but the exercise is designed to do a number of things.

First, it is designed to let you experience what uninterrupted writing feels like. After all, for many people that feeling is uncommon enough that it is worth a special effort to achieve it.

Second, the rules of freewriting force you to view the words flowing from your pen less critically, thereby letting you generate more writing per unit time. How much did you generate in your first 6-minute exercise? Many first-time freewriters generate between one-half page and one and one-half pages (single spaced) in 6 minutes. Although freewriting is not directed at a particular task, it does give you some quantitative feeling for your potential to generate lots of material in a short time. To extrapolate your experience, simply multiply your 6-minute page count by 10 to estimate your freewriting productivity on an hourly basis. For the average first-time freewriter this translates to something like 5 to 15 pages per hour. Even allowing for large amounts of wasted material, a writing process that allows you to move forward in a relatively uninhibited way has the potential for greatly increasing your writing productivity.

Third, and perhaps most important, freewriting is designed to give your writing a qualitative boost. Released from the bonds of judgment, you are able to generate new notions, partially formed ideas that previously would have been stillborn because they didn't come out perfect in wording, grammar, and usage. Thus, stripping away the tendency to be so self-critical can result in a step change in quality, the exploration of new rhetorical devices, and the discovery of what is often called a writer's voice.

Of course, all these things are possible, but one session does not a freewriter make. A sensible way toward improvement is to buy a notebook dedicated to your freewriting activities and to simply sit down at a designated time every day for a month, or two, or three, or six, and just freewrite. If you can get into the habit over a prolonged period of time, you will be able to follow the ebb and flow of your thinking and to play with words in a constructive way with no downside risk of criticism and with much upside potential for becoming a better, more productive writer. As you do more freewriting, I encourage you to play with variants of the exercise. A particularly practical one is to turn your writing consciously toward a specific topic. Exercises along these lines will loosen your writing so you'll be better able to create some required piece of writing. As you turn away from freewriting and move toward more directed writing tasks, you'll find that you need to modify the technique somewhat to bring it down to earth, but many of the lessons of freewriting will transfer quite easily to the directed task.

5.3.2 Directed Writing for the Real World

The undirected exercise of freewriting helps loosen constipated pens but by itself is less than helpful in getting a needed piece of writing done. For that, you need a plan of attack, an approach to writing that will let you finish that memo, that section, that chapter, or that report in a timely fashion. In this section, we take the lesson of freewriting, modify it so we can direct the course of idea generation, and use it in conjunction with a specific approach to revision. We will call this approach *directed writing*, and the process is designed to give busy engineers an everyday approach to getting important writing tasks done well and quickly.

Directed writing consists of three components:

1. Quickplanning
2. Directed creation
3. Cut-and-paste revision

Taken together, the three elements form a comprehensive approach to getting writing done quickly and well.

Quickplanning

No mention has been made of the writing teacher's perennially favorite tool: the *outline*. This omission has been intentional; I almost hate to give outlining any role in this process because it is so often misused. In a sense, outlining is an attempt to figure out what you're thinking about before you think it, and as we've seen in freewriting, we are perfectly capable of getting our thoughts on paper without the rigid discipline of lists. The dangers in constraining yourself to a rigid outline are twofold.

First, an outline prevents exploration of ideas as you think them; because you've got an outline, you will be tempted to follow it in serial fashion. Unfortunately (for outline-based writing processes), the human mind is not a serial computer. It works by association, and when an idea pops into your head, it is often worth exploring that idea right then and there, lest you lose the thought forever.

Second, the outline often misses important interrelationships between entries, interrelationships that should determine the ultimate ordering of the material. But, once again, if you've begun with an outline, you will be tempted (for reasons of pride, inertia, or sloth) to follow that original plan.

Thus, it is easy to conclude that a rigid outlining procedure can be as much of a hindrance to good writing as a help. On the other hand, many writers—myself included—feel uncomfortable writing without some sense of where they are going. That is why I recommend and use a coarse outlining scheme I call *quick-planning*.

To get the right mental image of quickplanning, think of the freewriting process; now, instead of writing down complete thoughts or sentences, imagine hopping around from one idea to the next, writing down only enough to convey each thought. As a mechanical device, it is useful to place a bullet—a large dot—in front of each idea fragment. Just write the thoughts down; don't worry about the order in which you generate them, and don't worry if the categories overlap. After you've written down the main points or ideas, you can massage the list into a preliminary ordering if you like, but if a good ordering isn't obvious, any ordering will do at this stage. You simply want quickplanning to give you mental keys that will help the *associative process* of generating raw text. I emphasize that this initial stage of planning should not be labored or lengthy. If you miss important points, you can always come back later and fill them in. You are simply trying to prime the associative pump for the next stage of directed writing: directed creation.

You may begin to see that the process suggested here is not as neat or as orderly as the usual one-shot, outline-and-write process that is often adopted. Of course, there is little that is orderly or neat about the human mind itself. That is not a criticism of our thought processes, however. Our minds are rich association machines that carry our thoughts along a synaptic superhighway of concepts, ideas, and relationships. One of the jobs of an appropriate writing process is to key into the way our minds work and to help us explore various outposts of thought along this neuronal road.

Directed Creation

With a rough quickplan of key topics before you, it is a straightforward matter to generate raw material for your piece. Again, you want to achieve a frame of mind similar to that of freewriting, except now you want to take away the "free" and direct the writing toward the quickplan form of bulleted topics. You should be careful to maintain the exploratory spirit of freewriting and not restrict yourself too much. If directed creation generates promising associations, why not go with them? You can even afford to be distracted by the occasional tangent (if you have the time) because sometimes by writing through tangential thoughts, you can discover something important about the topics of primary interest, things that would not have surfaced otherwise.

If we keep these things in mind, we can define the process of directed creation by seven rules:

1. Write associatively, using the quickplan bulleted topics as a guide until your thoughts run dry or until the time budgeted for each topic runs out.

2. Do not stop and, insofar as it is possible, do not lift pen from paper.

3. Do not cross out any writing, not one word.

4. Do not worry about spelling, grammar, punctuation, or structure.

5. Stick to the point somewhat, allowing some time to develop related topics and tangents.

6. If you become stuck, move on to the next bulleted topic or shift to a thought generated elsewhere in the process of directed creation.

7. Write on every other line, and do not write on the back side of a page.

These rules are similar to those of freewriting, with a number of modifications and the addition of one other.

The changes to rules 1 and 5 reflect the primary modification to freewriting necessary to achieve a more directed flow. Again, the bulleted topics developed during the quickplanning session should be used as mere seeds for the larger piece; in general, be more trusting of what's in your head, and follow your writing where it leads. If you are writing to a deadline, you can cut off tangents that don't seem germane and allocate fixed amounts of time to generate writing on a particular topic—but some freedom should still be given to pursue thoughts as they come.

Rules 2, 3, and 4 remain largely intact; it is particularly important to follow rules 2 and 3, to keep the writing moving forward without crossing out.

Rule 6 has been modified to recognize that indeed you are now writing about something. If you should get stuck (or run out of things to say about a particular topic), you can simply move on to the next bulleted item. Of course, when the spirit moves you, you should not hesitate to move back to a partially treated item.

Rule 7 has been added to allow for the addition of snippets of writing during the process of revision. Rule 7 recognizes two things: (1) It is useful to have extra space between lines so you can add or *interpolate* missing material, and (2) it is inconvenient to paste one page to another when material is written on the back of the sheet being so affixed.

Together, the seven rules form a basis for directed creation. The proof is in the doing, which leads us directly to an important practice exercise.

Exploration Exercise

Quickplan and perform directed creation for a two- to four-page, double-spaced, handwritten essay on the topic "What Are My Aspirations and Goals for the Next Five Years?"

In answering, try to be concrete. Generalities and platitudes are not helpful in exploring what you think, nor do they make for very lively writing. Also, don't spend much time apologizing for your lack of certainty. It is hard to project what will happen tomorrow, let alone what will happen in 5 years. On the other hand, a vision and goals are useful in evaluating opportunities and making decisions. Remember that it is unnecessary to force your writing to be too coherent at this point; you will have the opportunity to reorganize, interpolate, and extrapolate when you perform cut-and-paste revision.

Aside on Writing by Hand versus by Computer

The widespread availability of word-processing software and personal computers raises important questions as to how to integrate these tools into one's writing. Although computers profoundly change the final preparation and presentation of a manuscript, they really have little to do with developing an effective process of writing. After all, the grayware between our ears is the most important contributor to an effective piece of writing, not the software we use for manuscript preparation and presentation or the hardware on our desks.

Having said this, I have to concede that in this day and age most manuscripts will be processed at some point on a computer. The appropriate question, then, is at what point should computers be introduced into the writing process? At one extreme, some might argue for introduction of word processing at the very start: Compose text directly at a computer keyboard. At the other extreme, some might argue for entry of text only following the completion of an edited manuscript. Here, I come down firmly for both extremes (and for some alternatives in the middle), depending upon the length and nature of the document and the level of experience and confidence of the writer.

Direct entry of text into a computer seems to offer the advantage of saving time, but there are two dangers associated with direct composition at keyboard: *nice-output syndrome* and *hurry-up-and-wait disease*. Nice-output syndrome occurs when an inexperienced writer looks at the nicely formatted output that comes off the laser printer and hesitates to revise the piece of writing. When words are neatly printed, there is a tendency to treat them with more respect than they deserve. After all, they are nicely printed and look pretty darn professional, don't they? But good writing involves a good bit of iteration. Anything that makes us reluctant to try different combinations or orderings is something to be avoided.

Hurry-up-and-wait disease occurs because a good touch typist can easily outrace the speed with which he or she can think (Knuth, Larrabee, & Roberts, 1989, p. 14):

> *Upon receiving a question from the audience concerning how many times he actually rewrites something, Don told us (part of) his usual rewrite sequence: His first copy is written in pencil. Some people compose at a terminal, but Don says, "The speed at which I write by hand is almost perfectly synchronized with the speed at which I think. I type faster than I think so I have to stop, and that interrupts the flow."*

In some ways, these suggestions for intentionally slowing down the writing process through the choice of a particular medium (e.g., pencil and paper) parallel the just-in-time methods of modern manufacturing that emphasize a steady line speed and well-synchronized operations. Note that the author of the passage quoted above is talking about fairly involved manuscripts, like journal articles and books, which are both novel and fairly long. In composing long, new material, many writers would do well to write raw text out in longhand to avoid the scourge of hurry-up-and-wait disease.

On the other hand, many business documents are neither novel nor long; instead, they are fairly short and on well-worn topics. In these cases, there is little risk in direct composition on a computer. For example, memos, letters, brief reports, and similar documents might not require much handwriting. Especially when you've written extensively about a subject, your thinking can keep up with your typing, so the hurry-up-and-wait disease is not as much of a concern. Also, for shorter documents, the amount of reordering possible is so limited as not to be a worry.

Having said these things, however, it is still important for all writers of documents, long or short, to have a good feel for a productive writing process. Therefore, for the exercises here, and for writers who are trying to improve, I recommend handwriting for text generation and physical cutting and pasting for revision. Even if you rarely actually use this mode of writing, the mental image of preparing your document in this way can only help you straighten out your logic and flow whatever medium you choose for recording your written words.

Cut-and-Paste Revision

Having practiced directed creation, you are now free to plow forward. By not constantly trying to get things right the first time, you should have little trouble generating fairly large numbers of pages of raw material. With this important capability under your belt, it is time to learn to put this raw material into a more organized form through cut-and-paste revision.

Cut-and-paste revision has a number of physical requirements:

1. A red pen
2. Scissors
3. A glue stick
4. Sheets of unruled paper

These items, together with a ruled writing pad and pen, form your cut-and-paste tool kit.

To actually begin the cut-and-paste process, first read through the raw output of your directed-creation session, cutting out those passages that contain something important, whether that something is a word, a phrase, or a paragraph. Then treat the assembled fragments like a puzzle, attempting to fit the pieces into an orderly sequence.

Of course, some of what you've already written will itself need revision. That's exactly why you skipped every other line, leaving room for interpolation.

Also, at this point, a red pen is your best weapon for slaying wordy mammoths that roam through your sentences and paragraphs.

As you piece the puzzle together, you'll want to make the draft more permanent by gluing the fragments to another sheet of paper. Blank computer paper works well for that purpose, but any letter-size sheet of paper will do. As you paste pieces together, you will notice awkward transitions and even whole subsections that need rewriting. There is no need to panic. Simply throw yourself into directed-creation mode, generating text to fill the current need; as you switch between modes, however, it is important to recognize the mode shift. Don't fall into the bad habit of trying to write any of the new material right the first time. Instead, when you click into directed-creation mode, obey all the rules of that game. When you've written enough stuff to fill in the missing blanks, switch back into revision mode, cutting, red-penning, and pasting to your heart's content.

Again, as we discussed earlier, depending upon the length and novelty of the document and the experience and confidence of the writer, it may be fine to do the cutting-and-pasting phase on a computer. Even so, it is good to have experienced real cutting and pasting, at least once. The sense of writing as experimentation comes about most easily when you actually do cut and paste little scraps of paper together. Having that mental image can be invaluable for the writer who prefers to compose and revise at the keyboard.

It has taken some time to describe the whole process, but now that we have, it's time to finish the exercise we started previously.

Exploration Exercise

Using the material generated in the previous practice exercise, perform cut-and-paste revision for a two- to- four-page, double-spaced, handwritten essay, "My Aspirations and Goals for the Next Five Years." Do not type the final document. Simply edit and complete the cut-and-paste copy by hand.

Having undergone the entire process once, you can begin to see the productivity gains and qualitative advantages of separating writing from revision. Of course, becoming a better writer is a never-ending quest. Understanding good writing habits and process is critical as is understanding common organizational devices and content.

5.4   GETTING THE CONTENT AND ORGANIZATION RIGHT

Developing good writing habits through freewriting and directed writing are critical skills, but understanding common conventions of organization and content are also important.

Writers of fiction often want to keep their readers in suspense. Shocking, surprising, humoring, or dismaying your readers in fiction is a virtue, but in technical and business writing, we must recognize a common characteristic of technical-business readers. They are all *busy*. Suspense, mystery, humor, or other emotional devices must usually give way to reading that is concise, to the point, and informative. Therefore, it is possible to prescribe certain devices of organization and style that will help the technical-business reader understand the material being presented more quickly. In particular, four elements of concise business writing are crucial:

1. BPR: background, purpose, and road map
2. Lists and amplification
3. Sectioning, titles, and headings
4. Summaries and conclusions (and knowing the difference)

If these elements are mastered, business writing can be more easily written and read.

5.4.1   The Primary Structure of Business Writing: BPR

Once the technical writer learns to just write, line after line can be generated with speed, but another problem focuses on the content: What should he or she write about? In detail, this will depend upon the writer's subject, but every piece of technical writing has three critical needs, needs that must be fulfilled at first and then over and over again in every piece of writing (Mathes & Stevenson, 1991).

Specifically every document, section, or major element of a technical writing must explain its:

1. Background
2. Purpose
3. Road map

At the beginning of a document—especially—and at the beginning of any section or subsection as well, the writer has a problem. At the beginning of a document, the writer faces a *sharp discontinuity* in a reader's understanding. Prior to the first sentence, the reader has little or no idea of the subject of the piece, its context, its purpose, or what is to follow. It is the writer's job to fill in the gaps quite speedily at the beginning of the document and in succeeding major sections. The first task, therefore, is for the writer to provide what has been called *background*. What is the motivation for and context of the document? What were the enabling events that led to the document's needing to be written? Who are the key players? Background is crucial, but it should be provided quickly, with only essential detail, because the clock is ticking until the writer explicitly states the purpose of the piece.

The *purpose* or *rhetorical purpose* of a piece is the objective of the *piece of writing* itself. Note that this is distinct from the purpose of the project or problem that the document may address. Sometimes students in presenting the rhetorical purpose beat around the bush or try to be subtle. Subtlety is for mystery novels, and in business writing it is better to simply come right out and blow the trumpet with phrases such as, "The purpose of this report is. . . " or "The goals of this memo are. . . . " Students sometimes learn to announce their intentions more subtly in freshman English, but business people are busy. They require clear landmarks in their reading, and the wise technical writer will give them what they need.

Following the background and purpose, it is time to give a *road map* to the remainder. At the beginning of a document, this road map foreshadows the subject's major topics in the remainder of the document. In a section or subsection an intermediate road map lays out the topics for that section or segment of writing.

Road maps are often left out by novice writers, and it shows. Giving a road map builds a *mental model* for the reader of what is to come. It gives the reader a preliminary point of view that helps create appropriate expectations for what is to follow. Without road maps the reader isn't sure where he is and he isn't sure where he's going, and from the reader's perspective, this is most disconcerting, like being lost in a fog. Yet, by simply handing the reader an occasional road map to the remainder, such discomfort can be avoided and greater clarity can be achieved.

Note that the BPR structure is iterated at *all* levels of a business or technical document. At the beginning of the document, BPR needs to be given completely and explicitly. Later, in subsequent sections and subsections, BPR can be given briefly or implicitly, but it still needs to be given. Business or technical readers often skim material looking for answers to their specific questions: BPR given properly throughout a document lets the reader constantly know where he or she is, thereby ensuring that the document will be understood and used properly.

5.4.2   Lists and Amplification: Technical Writer's Best Friend

A terrific way to give road maps or otherwise help organize a piece of writing is by using the numbered or bulleted *list*. The list clearly announces the forthcoming structure, thereby alerting the reader to coming attractions. The list has two benefits:

1. It announces the items to the reader in an easy-to-grasp manner.
2. It simplifies subsequent writing by encouraging *amplification*.

Think of a list as a rough outline, and the presentation of a list acts as an automatic type of road map to coming attractions. The mental model is especially easy to grasp as it is laid out in numbered or bulleted points.

Moreover, giving a list simplifies the writing task considerably because all the writer needs to do is write one paragraph (section, subsection, etc.) for each bulleted or numbered point, and the organization and content of the document go hand in hand. Go back through *this* book and look at how lists and amplification were used to first highlight a road map and then fill in the blanks.

Newsletter Format   A very effective way to organize short memos and e-mails is in *newsletter* format. Create a series of headlines, which are set in **bold**, *italic*, CAPS, or other fonts that stand out, and then complete the paragraph with an amplification of the topic headline. A series of newsletter paragraphs can be proceeded by a list, or the headlines themselves can function as a list if the document is short enough. Either way, busy people like this style because it allows them to skim through to the part that really interests them.

5.4.3 Sectioning, Titles, and Headings

When I coach students writing long technical reports for a senior design project or business plan course, I find that there is a tendency to write overly long sections that are not broken up into subsections. This is a mistake because it makes it harder for busy business-technical readers to skim the text and find what they need. A key element of good business writing is to break up documents into reasonably sized pieces with appropriately chosen headings or titles.

The art of choosing titles for a document or headings for a subsection may be best understood if we recognize two key dimensions or characteristics of titles or headings:

1. Informative
2. Interest provoking

Titles or headings can be either purely informative, purely interest provoking, or some combination of both. In straight technical writing it is best to err on the side of being informative, although from time to time, generating interest by using a title or heading that is somewhat unusual can be a way of keeping your reader engaged in reading the document.

5.4.4 Summaries, Conclusions, and Distinguishing the Difference

Two elements are needed at the end of a piece of writing, and they are usually covered under a pair of headings:

1. Summary
2. Conclusions

There is a good bit of confusion between the two, and here we examine each one and make key distinctions between them.

A summary is concisely summarized in the advice offered to Army officers in writing their memos:

Tell them what you're going to say,
say it,
then tell them what you said.

A summary is the "Tell them what you said" part of the closing, and indeed it is critical to integrating the piece of writing in your reader's mind. In a sense, a summary is a backward looking road map where you revisit the milestones of the piece to tie everything together. Just as a road map creates a mental model of what is coming, the summary concisely refreshes the reader's mind of what was said. By abstracting the essence of a complex argument, the writer encourages the reader to remember key elements of the discussion.

Conclusions are distinct from the summary and to understand them, we recognize that they answer a critical question for the reader:

> **Conclusions answer this question:** *How should the reader's thoughts or actions change as a result of having read the piece of writing?*

In a sense, conclusions are a call to *think* or *act* differently as a result of the material presented. Conclusions that involve largely changes in action are sometimes given the special name "recommendations." But whatever you call them, conclusions and recommendations are important because they draw out the *consequences* of what was said for the reader.

5.5   EDIFYING EDITING

Separating writing and revision through the use of directed writing and cut-and-paste revision can be such a liberating experience that the recently initiated may get the idea that the first products of these efforts need little or no additional attention. This idea is usually a mistaken one; most drafts need a number of passes of careful editing before they are ready to meet their final audience.

Much classroom time in grade school, high school, and freshmen English is spent on matters of grammar, punctuation, and sentence structure. These lessons need to be applied in the editing process with a vengeance. Less time is spent in these classes on matters of style. There are a number of editing maxims that can help enliven your writing:

1. Omit needless words.
2. Use active voice.
3. Be specific.
4. Enhance parallel structure.
5. Enhance sentence cadence and rhythm.

This list is far from exhaustive, but these elements of style are among the most important.

In the absence of any other stylistic guidance, we can do much worse than to follow Will Strunk's advice and "omit needless words" (Strunk & White, 2000). While it is something of an exaggeration to say that all good writing is brief, much bad prose is wrapped in layer upon layer of verbal fat, and the dietician's red pen is often in order. Good examples of vigorous and not-so-vigorous writing can be found in *Elements of Style* by Strunk and White (2000).

Active voice is essential to vigorous writing, and it is in this regard that so much technical writing falls flat on its face. Some of the stumbling is self-inflicted through adherence to the old saw "Never write in the first person." While it is true that it would be unwise to adopt the somewhat breezy style of this text in a typical consulting report, I doubt whether Western commerce would come to a screeching halt if the collective "we" or the individual "I" were used in more business writing. This single step immediately forces writing to become more active and lively. Even if you are required to write in the third person, it is possible to enliven your writing by avoiding anonymous subjects and talking about project teams, projects, or experiments more directly. For example, we might write one of the two following active sentences

- In this report, I (we) examine the X-ray hypothesis. [*Writer as subject*]
- This report examines the X-ray hypothesis. [*Report/project/team as subject*]

as alternatives to the passive construct

- In this report, the X-ray hypothesis is examined. [*Topic as subject*]

Active forms are more direct and lively, and oftentimes they are shorter.

One common flaw is the substitution of a high-minded, general-sounding word or phrase when a simple word would do. Buzzwords such as "areas" or "issues" will wrongfully take the place of concrete words such as "tasks" or "problems." Fuzzy-headed verbs such as "involve" or "consider" will be used in place of action words that actually describe what is taking place. Concrete, specific language paints a picture in the reader's mind that lasts because it engages active thought.

Pursuit of parallel structure is an important principle that is often ignored by the less experienced writer. For example, suppose the listing of topics for this section had been written as follows:

1. Omit needless words.
2. Use active voice.
3. Specific words are helpful.
4. Enhance parallel structure.
5. Rhythms should be watched.

Each phrase is grammatically correct, but items 3 and 5 break the parallel structure set up in numbers 1, 2, and 5. The verb–noun order established in the three is upset by the noun–verb order of 3 and 5. The change in pattern is mentally disruptive and ultimately prevents the writing from being as vigorous as it could be. Existing parallel structure should be identified and enhanced; building parallel constructs can bring an immediate power boost to your writing.

Attention to rhythm or meter is most often associated with poetry, but good cadence in prose is no vice. (Prose is poetry that can earn a living.) In fact, many stylistic rules relate to an attempt to achieve better meter or rhythm in prose writing. For example, a common error of the inexperienced writer is to string together sentences of similar length and structure—the see-Spot-run, run-Spot-run syndrome. Although such sentences are grammatically correct, the error is one of boringly repetitious structure.

Although there is no foolproof way to recognize and correct bad cadence, it is useful to read your writing out loud. Once you identify passages with awkward meter, they sometimes can be fixed through the joining that comes with the use of a conjunction, the splitting that comes from the use of a semicolon, or the careful rearranging that achieves well-modulated structure. Though at times more major surgery is necessary, reading your writing aloud and listening to its rhythm can help improve its power and impact.

5.6   IMPROVING YOUR WRITING

Good writing is a journey, not a destination, and there are a number of side excursions worth the fare:

- Reading more
- Writing more
- Getting professional editorial help

We briefly examine each of these possibilities.

One of the easiest ways to improve one's writing is to read more, and one of the easiest ways to do that on a regular basis is to read a first-rate newspaper. If you don't already take a regular paper (and even if you do), consider subscribing to and reading *The Wall Street Journal*. It has some of the best writing of any major national newspaper. Moreover, the style that its editors have adopted—far from being the turgid business prose you might imagine if you've never read the WSJ—is some of the most lively and engaging newspaper writing you will find anywhere. Moreover, with so many business leaders getting their daily fix of the WSJ, you can do much worse than to use that paper's prose as a model for improvement.

Good nontechnical, nonfiction writing can provide a model for exposition. Well-written history and biography (e.g., works by H. W. Brand, Martin Gilbert, Paul Johnson, and William Manchester) are useful for learning to handle time and sequence. Well-written popular science books and novels (e.g., by such authors as Isaac Asimov, Michael Crichton, Steven Levy, and James Gleick) are useful in learning that the presentation of factual information need not be dull.

Much has been written about writing, and engaging the literature of writing can be helpful, but it is important to pick books that ring your truth bell. As has been mentioned, Peter Elbow's *Writing with Power* (1998) is a good starting place if you are interested in learning more about writing as a process; *The Elements of Style* by Strunk and White (2000) is an amazing little volume on almost everyone's short list of writing books. For matters of grammar and punctuation, I recommend Karen Gordon's *The Deluxe Transitive Vampire* (1993a) and *The New Well-Tempered Sentence* (1993b). These clever books cover difficult material using quirky, humorous examples and counterexamples. For a fairly encyclopedic view of almost all matters of detail and form, there are few better sources than *The Chicago Manual of Style* (University of Chicago Press, 2003).

Another easy way to improve your writing is simply to write more. Shortly after I joined the mechanics faculty at the University of Alabama, I was "volunteered" to be the secretary of the weekly faculty meetings. Rather than gripe, I took that opportunity to develop my writing skills (and occasionally to tickle my colleagues' funny bones), and along the way I learned a good bit about conveying mundane details in an interesting way. Similar opportunities pop up from time to time for everyone, whether at work, in a civic or church organization, or simply in writing letters to friends and relatives. The prevalence of electronic mail (e-mail) and blogs are opportunities to try your hand at informal pieces of writing from time to time. I urge you to view these and other writing episodes as an opportunity to improve.

If you have the opportunity to get professional editorial help, take it. Having a "pro" go through your work helps you identify your most common errors of grammar, punctuation, and usage. A good editor can also help with sentence, paragraph, and overall structure. English departments at universities and community colleges can often put you in touch with a qualified editor, but it's a good idea to get references from writers who have worked with the individual before making any commitments.

SUMMARY

In this chapter, we have considered why many engineers dislike writing and how to combat this difficulty by articulating a straightforward approach to the process and content of business-technical writing.

We've seen how the criticism of early writing efforts often leaves engineers uncomfortable with their writing ability. This lack of confidence manifests itself in a herky-jerky mode of writing that involves a little writing, a lot of crossing off, and a thorough familiarity with the location of your wastebasket. To learn to overcome this start–stop mode of writing, this chapter has considered an exercise called freewriting, which permits us to explore our thoughts without fear of criticism as we write with no particular purpose in mind. We have then seen how to bend the process of freewriting to a more directed piece and recognized how bouts of directed writing followed by cut-and-paste revision can be a fairly effective means of completing the writing tasks before us.

The chapter has also considered certain patterns or devices in business-technical writing that simplify the writing task. The most important of these has been called BPR, or background–purpose–road map. At the beginning of every e-mail, letter, memo, and report we start with crucial background material to help the reader understand the context of the written piece. Thereafter, the rhetorical purpose informs the reader why the piece was written, and the road map builds a mental model of the remainder. Other devices such as lists and their amplification as well as the liberal use of headings and subheadings are useful in technical writing and make the task that much easier. Knowing the difference between summaries and conclusions and using them are helpful in closing a written piece.

We've also seen in the chapter how "omitting needless words" and a number of other useful rules of editing can help finish a piece of writing and make it easier to be read and understood.

For many entrepreneurial engineers, following these steps can help unlock the writer inside. The steps require practice. At first it is difficult to avoid being critical during the creative phase of the work, and then it is difficult to master BPR and the other secrets of creating compelling content, but, with practice and time on task, the dividend can be nothing less than a lifetime of productive and confident writing.

EXERCISES

1. Freewrite for 10 minutes every day for 2 weeks. In a short paragraph, compare and contrast the quantity and quality of your freewriting on the first and last days of the trial period.

2. Use the directed-writing process to write a brief essay, taking either the affirmative or negative position, on one of the following topics:
   - Engineers are properly appreciated in society.
   - Engineers do not know enough about nontechnical topics.
   - Laypeople do not know enough about technical topics.
   - The United States' competitive position in the global economy is declining.
   - American executives today are unethical relative to those of past years.

3. Write a brief essay arguing the opposite side of the issue you chose in Exercise 2.

4. Write a letter seeking employment with a company that interests you.

5. Write a thank-you letter to an individual who interviewed you for a job at a company.

6. Write a short autobiography.

7. Write a resume.

8. Write a short family history tracing some members of your family back at least two generations.

9. Write a brief biography of an interesting family member or an acquaintance.

10. Write a set of instructions for a game, a machine, a piece of software, or other complex entity used by human beings.

11. Write a brief description of the principles of operation of a device, process, or algorithm.

12. Write a brief paper describing the solution of a calculus or physics problem.

13. Write a brief how-to article on some aspect of a favorite sport or hobby.

14. Write an eyewitness account of the most unusual social gathering you ever attended.

15. Write copy for a brief marketing brochure for a technical product or service with which you are familiar.

16. Form a group of five writers and choose a topic from the list in Exercise 2. Have a brief group quickplanning session and then divide the bulleted items among the group members. Each member should then perform cut-and-paste revision using the raw material of the entire group. Exchange results; discuss the differences of organization and style represented by the five essays.

17. With another writer, select a topic from the list in Exercise 2. One of you should do directed creation for the affirmative, the other for the negative. After both of you have completed your initial writing, each writer should take both sets of raw material and generate a balanced essay that examines both sides of the chosen issue. Discuss similarities and differences in your final essays.

18. Buy a copy of Strunk and White (2000) for your library Read it. Make a list of your top-five writing errors and work on correcting them. Write an essay on your experience in so doing.

19. Take minutes for a meeting of an organization or committee. Write them up in newsletter style. Gage the reaction of your organization-committee members to this style of minute taking and write a short essay about your experience.

20. Start a personal, academic, or technical blog and make at least one short entry in the blog every day for a week. Write a short essay about your writing experience and the ways in which having a public audience affected (or not) your writing style.

Chapter 6

Present, Don't Speak

6.1 SPEECHES VERSUS PRESENTATIONS

It seems that we're always in the middle of an election. And elections mean speeches: stump speeches, impromptu speeches, TV speeches, speeches after dinner, speeches in schools, and speeches in malls. With the prevalence of stand-up public speaking in the political arena, it would be easy for the new engineer to become confused and assume that the same mode is commonplace in business communication. The truth is that it is the rare businessperson today who gives a stand-up speech, at least for everyday business matters. With the advent of overhead projectors and transparencies, businesspeople stopped giving speeches and started making presentations: talks supported by projected visual material. The advent of laptop computers, LCD (liquid-crystal display) projectors, and powerful presentation software has advanced this trend.

In this chapter, we'll examine the reasons why you should join this mass movement and learn to prepare and deliver effective presentations. Specifically, we'll examine the reasons why you should present—not speechify—and we'll consider the elements and the process of presentation preparation. We'll also consider methods for preparing effective transparencies or PowerPoint presentation files and ways to sharpen your delivery skills.

6.2 WHY PRESENT?

After you have some experience giving transparency-based or PowerPoint presentations, the reasons for presenting rather than speaking will seem almost self-evident. That this knowledge is not genetically transmitted was brought home to me several years ago when I was a group advisor for a Senior Design Project at the University of Illinois. The design team—a talented group of motivated senior engineering students—had prepared their spiel and wanted to rehearse in front of me, so I could give them feedback. As soon as they began to talk, I realized—to my amazement and chagrin—that they were giving a speech. Sure, they had a transparency or two as window dressing, but it was a speech

The Entrepreneurial Engineer, by David E. Goldberg
Copyright © 2006 John Wiley & Sons, Inc.

they delivered, with one student working from notecards, one reading verbatim from a script, and one working from (a faulty) memory. After listening for a few minutes, I asked them why they were making their lives so difficult. Uniformly, they answered that they thought that this was what the big guys did and that this was what they were expected to do for their first shot at the big time. Fortunately, they were persuaded that they could be more relaxed—and more communicative—if they would present, not speak. They went on to give a solid presentation, but the experience left me with the impression that there is a wide gap between what students think a business presentation is and what such presentations really are.

Formal oral communication in business is now dominated by transparency-based or PowerPoint presentations for two good reasons:

1. Presenting is easier to do.

2. Presenting conveys more information to your audience.

One of the reasons I couldn't believe my students chose to make a speech was because I know how hard it is to make a good one. Consider the three ways you can give a formal speech. You can memorize the speech—but for most people it is too easy to forget portions and become flustered. You can read from a script—but, unless you are a good actor, it is very hard to read and not sound as though you are reading. Or you can work from notecards (and this is perhaps the most sensible way for the occasional speech maker to work)—but even then, because your audience is not visually occupied, they will focus more on you, your mistakes, and your use of notecards. The presenter needs no such extra attention; nor does he or she need the stress it can cause.

Contrast the difficulties of making a speech with the utter ease of giving a well-planned presentation. You walk up to your laptop and LCD projector, madly click your mouse, read the cues on each page of the presentation, and verbally fill in the blanks. And while you're doing your thing, your audience isn't minding you much. They are happily engaged with the material you are putting before their eyes, the same material that is providing you with cues to continue your talk. In this way, presenting is much more forgiving to the communicator, providing notes and props to help get through the talk.

Presenting is also advantageous to the audience. Well-planned transparencies or PowerPoint slides provide a second channel of information, augmenting the primary source—the speaker's voice. An audience member who misses a point from one source can often pick it up from the other. Moreover, in this video age it is not irrelevant that a transparency presentation is a visually engaging activity. An audience raised on television expects to have its visual field filled. A communicator who misses or misunderstands such an important audience expectation is simply asking for trouble.

Thus we are drawn to an interesting conclusion. When you have a choice, choose to present. You'll be more relaxed, the audience will be better engaged with the material, and more information will be conveyed.

6.3   PREPARATION MAKES THE PRESENTATION

That a presentation is easier to deliver than a speech certainly does suggest that we should choose to present whenever we have a choice, but ease of delivery does not imply that a presentation requires less preparation than a speech. In fact, because of the need for transparencies or Powerpoint slides, a good presentation usually requires more time for preparation than a comparable speech. Careful presentation preparation requires a number of key steps: audience analysis, subject selection, elements of a presentation, preparation process, and transparency design and preparation.

6.3.1   Audience Analysis

The best presentations come from serious consideration of *audience*. Experienced writers and speakers have a gut feel for their audiences and constantly adjust to audience needs, but the less experienced communicator has less of a feel and has to give more consideration to the approach chosen. To aid this process, it is helpful to prepare a short *audience brief* to help guide presentation planning for a particular audience.

In preparing the brief, there are three audience characteristics to keep in mind:

1. Motivation
2. Patience level
3. Educational and technical background

By far the most important of these characteristics is audience motivation. Why is your audience bothering to sit there and listen to you? Are they simply scratching an intellectual itch, or do they need to learn something fairly specific? Are most audience members there for the same reasons, or are different audience members there for different reasons? It is important to address both the motives and the variance in motives as you contemplate your target audience. Of course, the main reason that you contemplate the audience is to connect your material to their motives. This permits you to establish the appropriate angle from which to present your subject matter. Any subject can be presented from various perspectives. For a relatively homogeneous audience you may select one particular point of view; for a more mixed audience it may be necessary to present multiple viewpoints to connect with differently motivated individuals. Only by understanding motives, and by focusing your subject material toward those motives, can you hope to reach your audience—and thereby accomplish your goals as a presenter.

Patience may be a virtue, but in business, time is money; many of your audience members will have severe limitations on how long they can sit still for your message. Therefore it is important to tailor the length of your talk to the level of patience (or, more often, impatience) of your audience. For example, a CEO (chief executive office) with nine appointments before lunch and a plane to catch has one patience level, and a project engineer with only his or her workstation on hold has another. If you're faced with a situation in which both high-level managers and project engineers compose your audience, it may make sense to split your talk into distinct management and technical briefings, and this is often done. On the other hand, there will be times when you can expect fairly uniform patience levels in your audience—at technical society conferences, for example. Even then, audience patience can become an issue, especially if your talk exceeds the allotted time.

Just as levels of patience and length of talk must be matched, so must audience background and the intellectual level of a presentation. Waxing eloquent about a set of differential equations in front of an audience of Realtors is likely to evoke thoughts of cost variation among leases (differ-rental equations). Of course, spending time explaining elementary differential equations to a group of Ph.D. physicists is equally nonsensical. Therefore, assessing your audience's technical and educational background is vital. Once again, the trickiest audiences are those with mixed backgrounds. In the worst cases, it may be best to divide if one wants to conquer the heterogeneous audience.

Considering these aspects of your audience will help you design your presentation appropriately. Table 6.1 summarizes the connections between audience characteristics and presentation consequences. Although this section has been fairly clinical in dissecting the components of audience analysis, the more appropriate mechanism for considering a particular audience is the holistic preparation of an audience brief.

The following exploration exercise asks you to prepare an audience brief for a presentation regarding your goals and aspirations. If this exercise is to be performed in class, you may want to do an analysis of that target audience; if you are doing the exercise on your own, choose an imaginary audience (e.g., parents, colleagues, or potential employers) and stick to that imagined audience.

Exploration Exercise

You will prepare a seven-transparency presentation on the topic "What Are My Goals and Aspirations for the Next Five Years?" (You may change the title if you like.) For this exercise, write a one- to two-page, double-spaced, typewritten audience brief that identifies the motivations, patience level, and background of your target audience.

Table 6.1   Audience Characteristics and Presentation Consequences

Audience characteristic	Presentation consequence
Motivation	Angle
Patience level	Length
Educational-technical background	Intellectual level

6.3.2   Subject Selection

Subject selection will, for us, be something of a short subject. Often a business presentation arises out of a particular organizational need, making subject selection something of a moot point; however, there is some room even within the confines of a predetermined subject to choose among various aspects to be included in your presentation. There will also be times when you are in greater control and can choose your subject fairly freely—for example, when you choose to make a presentation at a technical conference or before a civic or church group. In any event, the prime directive of subject selection may be stated quite simply:

> *Within the constraints of organizational need and audience characteristics,*
> *choose a subject about which you are both knowledgeable and enthusiastic.*

Perhaps this seems like simply stating the obvious, but how many times have you seen presenters choose to speak when they knew little about a topic for which they had not very much enthusiasm? Audiences can see through an amateur in a New York minute—if not during a talk, then certainly during the question-and-answer period. Audiences are also fairly savvy at detecting whether a speaker has enthusiasm for his or her topic. The best reason to choose a topic is your passion for that topic. Good presentations don't grow on trees, and your enthusiasm will carry you through as you plan, prepare, and deliver your talk.

6.3.3   Elements of a Presentation

A good presentation sets the stage in the listener's mind, presents the core material, and sums up and spells out the consequences of what was said.

A simple structure that accomplishes these things contains the following elements:

1. Title
2. Foreword
3. Overview
4. Body
5. Summary and conclusions

Each of these must be considered in some detail.

Title

It may be somewhat unusual to think of the title of a presentation as a separate element, but a good title can be critical to the success of a presentation. It is the first element that the audience sees or hears; it is important both for creating a positive first impression and for building in your audience the desire to hear more. To do these things, a good title should be *informative*, reflecting

the material contained in the presentation, and *interest provoking*, creating desire
and anticipation.

Selecting titles is a somewhat mysterious art, and rather than be too analytical
about it, let's examine a sampler of actual presentation titles, ranging from the
fairly straightforward to the more purely provocative:

1. A Comparative Analysis of Selection Methods Used in Genetic Algorithms
2. Genetic Algorithms, Noise, and the Sizing of Populations
3. A Gentle Introduction to Genetic Algorithms
4. Six Ways to GA Happiness
5. Don't Worry, Be Messy

By the way, genetic algorithms (GAs) are search procedures based on the mechanics of natural selection and natural genetics (Goldberg, 1989). From an engineering perspective they may be used as optimization procedures, and they also have
something to say about an engineering theory of design, innovation, and invention (Goldberg, 2002). From a title–design point of view, these five titles run the
gamut from fully informative to fully provocative.

The first specimen is about as straightforward and descriptive a title as one
could have. Although it is a little long, it does convey a compact version of the
talk's contents. Moreover, there is little in the title to offend anyone; a title of
this sort is useful when a "Joe Friday" approach is called for ("just the facts,
ma'am").

"Genetic Algorithms, Noise, and the Sizing of Populations" is another fairly
descriptive title; notice, however, how the use of a *triple of topics* conveys the
breadth of the presentation at the same time that it creates wonder in the reader's
mind about how the three topics interrelate. Triples can be overused, but they are
an effective device if the juxtaposition is both informative and interest provoking
without being too exotic.

The third specimen illustrates how a straightforward title can be made more
interest provoking by the injection of a single word. "An Introduction to Genetic
Algorithms" would be a fairly informative, if pedestrian, title. The addition of
the single offbeat (and alliterative) word "Gentle" is enough to make the title
more inviting. Engineers need to approach the offbeat with caution, however.
Your employers may be more comfortable thinking of you as a serious engineer,
and it is possible to be too cute. Such matters are tricky, and all I can recommend
is that you develop your own good judgment.

The fourth example, with its "six ways," leans even more toward provocation
while still being reasonably informative. The actual presentation for which this
served as a title is about the use of six elements of practical GA theory to make
genetic algorithms work better in applications, and this title does hint at that,
creating interest by shrouding the six ways in mystery. (If you're interested in
drawing an audience to a presentation of theory, some mystery in one's title is
essential.)

The last title, "Don't Worry, Be Messy," goes almost all the way toward provocative at the expense of being informative. The presentation for which I actually used this title combines the material from two separate presentations that were more conventionally named: "Messy Genetic Algorithms: Motivation, Analysis, and First Results" and "Messy Genetic Algorithms: Studies in Mixed Size and Scale." I dared to use such an uninformative title because my audience at the International Conference on Genetic Algorithms was familiar with my work in this area, and I thought that the offbeat title might draw attention to what was essentially a review of material originally presented elsewhere.

As you can see, there is quite a bit of latitude that can be taken in designing an informative, interest-provoking title. These same principles can be helpful in choosing headlines, section titles, and other short, pithy descriptions that are used in presentations and written work as landmarks to inform or keep your reader interested. As you make more presentations, you will become more proficient picking effective titles and headlines. As with all elements of presentation design, if you start from knowledge of your audience, you will not go far wrong.

Foreword: A Word at the Fore

The foreword is an oft-neglected element of a presentation. I use *foreword* (not *forward*) to mean an element that sets the stage for a presentation on the larger scheme of things. The term *motivation* is sometimes used to characterize this important presentation element. Specifically, a foreword (or motivation) should contain two elements: background and rhetorical purpose.

Background creates context for a talk. What were the critical events or factors that led to this presentation? In what key ways is this talk necessary? After providing background, it is time to blow the trumpets and give the rhetorical purpose of the presentation. Phrases such as "The goal of this talk is ..." or "The purpose of this presentation is ..." announce the coming of the rhetorical purpose; the presenter should not be afraid to state what that purpose is. It is important, however, to separate the rhetorical purpose of the presentation from the goals or objectives of the project or the underlying work—they are not the same. For example, a long-term design project may have the project goal of designing a particular gizmo, whereas a project progress presentation might have the rhetorical purpose of examining specific accomplishments since the last report so that team members depending on the design can adjust or adapt their plans accordingly. Thus the rhetorical purpose has more to do with the expected consequences of the presentation than the expected consequences of the underlying work or project.

The foreword can be as simple as a brief statement made while the title slide is on the projector, or it can be a more involved statement accompanied by a more detailed sequence of slides.

Overview

Have you ever listened to a speaker who didn't tell you where the talk was going? More often than not, when the speaker finished you didn't know where he or she had been. One of the most important elements of a talk is the overview. It should provide a fairly clear road map for the talk: where it starts, where it twists and turns, and where it will end up. Often a single slide will suffice, yet the inclusion of that one slide will do more to help your audience than almost any other. A simple rhetorical device that works well (if it is not overused) is to repeat the overview slide between each major segment of the talk, highlighting the topic you are about to begin. This technique works best in talks where the subtopics are fairly independent. Whether or not you choose to update the route map in this fashion, you should always have an overview slide somewhere early in the presentation; not including it is an invitation to disaster.

Body

The body is, of course, the meat of the presentation, but it is difficult to say much about it in general, other than that the process of generating a good presentation body is almost identical to the process of developing good writing. One key to writing a good body is to divide it into elements and to give intermediate overview slides for each element of the body. A key to giving an effective presentation is to successively build mental models for your listeners of what is coming and then to fulfill the expectations that you have created. Good presenters are sympathetic to their listeners' plight. They know the average listener is always on the verge of being lost, and with that in mind they provide ample road maps and intellectual landmarks along the way so the listener can stay with the presentation.

Summaries and Conclusions

After you tell them what you're going to say and say it, you do need to tell them what you said. Specifically, you should do two things. You must *summarize* the key points of your talk and draw *conclusions* from the work. I have found that there is much confusion among my students regarding the difference between summaries and sets of conclusions. A summary is a simple recapitulation of the key points made during the presentation. Conclusions are those *consequences* for, or changes to, the state of knowledge or the state of the world that are a result of the work presented. In a practical sense, summaries are memory refreshers and conclusions are calls to action (at least calls to changing one's mind). Both are necessary, and both should be presented at the end of the typical presentation.

With the basic elements of a presentation on the table, it is important to examine the overall process of presentation preparation.

6.3.4 Preparation Process

Presentation preparation is so closely tied to the writing process that you'll find it is much easier to prepare a good presentation after you've written something on your intended subject. (It is also easier to write after you have made a presentation on a topic, but we must climb on this merry-go-round somewhere.) Writing before presenting does two things. First, it forces you to come to grips with the order of presentation and the transitions between topics. Second, the act of writing programs your tongue for talking. After you've turned a phrase or two on a piece of paper, standing up and giving a talk becomes much easier. Therefore, I usually recommend that my students write *something* before they present.

What you write will depend on whether the presentation is derivative or independent. By a *derivative* presentation, I mean a presentation derived from a piece of writing. Many of my presentations come after I have written fairly extensively on my subject. In such cases, the need for additional writing is limited to the compilation of a list of the topics chosen for inclusion in the presentation and the composition of a paragraph or two on any new topics not previously explored in writing.

By an *independent* presentation, I mean a presentation on a topic that you have not previously written about (or presented). In that case, it can be important to go through the mental processes of writing and revision by working up a piece the length of an extended abstract (two to six pages). By doing this as part of the presentation preparation process, you work through questions of topic selection and ordering fairly fully, and you do enough phrase turning to get some useful tongue programming.

Some may find it sufficient to work up an outline as an alternative to writing an extended abstract. Detailed outlines can inhibit creativity and idea exploration; however, once the subject matter is fairly fixed in mind, outlining a presentation should not be too risky. Nonetheless, should you find that the ideas are not as fully developed as you thought they were (as evidenced, e.g., by repeated periods of writer's block), return to a fuller exploration of the presentation's flow by writing more on your intended subject.

6.3.5 Transparency Design and Preparation

Once you have a feel for the flow of a presentation, you are in a position to prepare the presentation copy and to actually produce the slide file. We consider both of these topics after first considering the somewhat peculiar language of transparency copy.

Transparency-Speak

Preparing slides or transparencies may be thought of as an exercise in writing headlines. In other words, the designer of transparencies, like telegraphers of old, must make every word count, sometimes at the expense of complete sentences and other conventions of grammar and usage.

To get in the mood to "speak transparency," I find it useful to grab *The Wall Street Journal* and skim articles by reading only the headlines, both the article and section headlines. If you do this, you will notice the elimination of many adverbs, the suppression of all but the most necessary adjectives, and the use of high-impact nouns and simple verb forms.

Preparing Presentation Copy

We are ready to tackle the actual preparation of presentation copy. Usually this involves the writing of a headline and from two to six bulleted topics per slide.

Rather than becoming overly analytical, why don't we do as we did with titles and look at some copy from a presentation entitled "A Gentle Introduction to Genetic Algorithms"?

Let's start with the overview slide copy:

h: Overview

b: Motivation

b: GA basics

b: GAs in search and optimization

b: Advanced operators

b: GAs in machine learning

Here I've used the shorthand h: to denote a headline and b: to denote a bullet. Overview slides are usually fairly easy to assemble, as they are simple lists of the main topics of the talk.

Since this is an introductory presentation, most audiences that hear it are unfamiliar with genetic algorithms, and the term itself must be defined. The copy for this defining slide is as follows:

h: What Is a Genetic Algorithm (GA)?

b: A genetic algorithm is an adaptation procedure based on the mechanics of natural genetics and natural selection.

b: GAs have 2 key components:

b: Survival of the fittest

b: Recombination

Note that only essential information is included and that only a modest amount of information is presented on a single slide.

Further along in the presentation it is important to explain how GAs work. This how-it-works section is preceded by an intermediate overview slide:

h: GA Basics

b: Differences—In what ways are GAs different from other search techniques?

b: Mechanics—How do they work?

b: Power—Why do they work?

Keeping the listener updated on the progress of a presentation is important, especially in longer presentations. It is also useful as a means to keep yourself on the straight and narrow.

In this same section, another slide gives a laundry list of four ways in which GAs are different from other search techniques. One of those ways is that they are a blind-search technique. The blindness slide copy is presented below:

h: Blind Search

b: Canonical search must reject problem specifics.

b: Treat problem as black box:

g: [*black-box graphic*]

The g: is used to denote a graphic element (we'll discuss graphics briefly in a moment). Note that because of the inclusion of a graphic element, the text has been kept to a minimum. Too much text on a graphical slide (or vice versa) can be fairly distracting, but you do need some text to help tweak necessary associations that allow the audience to think the right thoughts as they listen and read (and allow you to say the right words as you speak).

Producing the Presentation

The technology of presentation production has changed dramatically in a short time. In the 30 years that I've been making presentations, I've used pen-stencil sets, pen-lettering devices, pressed lettering, strip-lettering machines, and computers. Of course, computer technology is the most flexible of the lot and is with us to stay. Therefore I recommend that you use some sort of word processor, graphic-arts package, or specialized presentation software. The PowerPoint program has become a de facto standard and has been used in the preparation of the slides shown here.

Again, rather than becoming overly analytical, let's just look at some sample slides created for the presentation discussed earlier. The first is the title slide (Figure 6.1). As you will notice, it contains the talk title, speaker name, address, and electronic address. Too many speakers begin their talks without mentioning who they are, where they are from, or what they will be talking about. It is a simple matter to ensure that your audience knows you, the title of your talk, and your affiliation.

Scanning this and the other transparencies (Figures 6.2 through 6.5), you'll see that a fairly Spartan PowerPoint style file has been used. Notice the clean design, simple graphic, and minimal use of different fonts. One of the unfortunate and unintended side effects of the explosion of computer graphics software has been to encourage a lot of junky graphics. The best policy for those who are not artists is to keep slide layout and graphics simple. The old acronym KISS (keep it simple, stupid) is appropriate here. The average presentation should be clean and clear; going overboard on graphics will probably have the unintended effect of persuading your boss that you've spent too much time on presenting and not enough time on solving the problem. Of course, if your line of work is computer graphics or visualization, then you will be expected to live up to standards of excellence established in your field. Otherwise, you overdo graphics at your peril.

Now it's time for the rubber—the PowerPoint—to meet the road and try some transparency preparation in the next practice exercise.

Figure 6.1 Title slide should contain the presentation title, the presenter's name, and the presenter's affiliation.

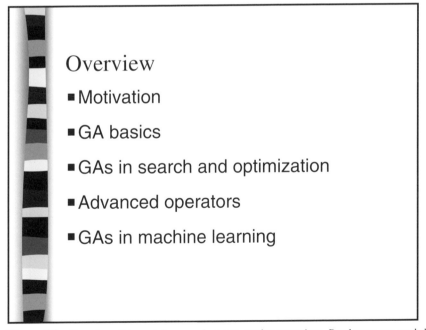

Figure 6.2 Overview slide contains a road map to coming attractions. Road maps are needed at the beginning of presentations and at the beginning of major sections and subsections.

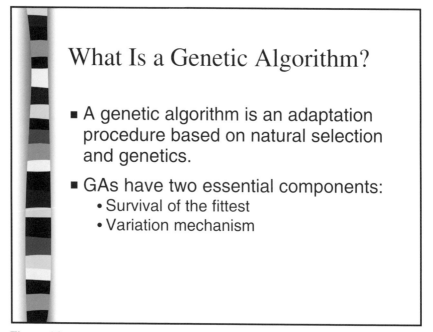

Figure 6.3 Definition slide gives a definition of a key term. Other body elements provide the primary information to be conveyed by the presentation.

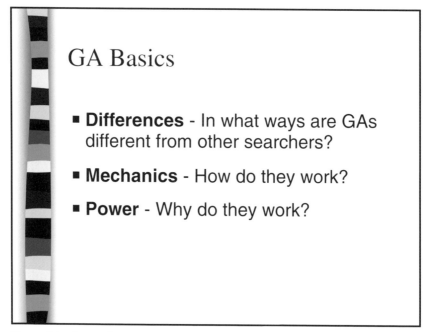

Figure 6.4 Intermediate overview slide gives a list of topics for a section. It is especially important in long presentations to update the route map.

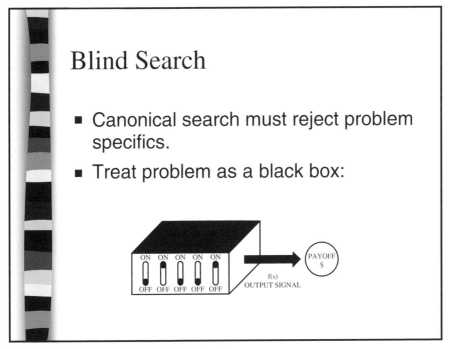

Figure 6.5 Sample topic slide with a graphic element shows text and graphics working together. Too much text on a slide or too busy of a graphic element can interfere with overall visual effect.

Exploration Exercise

Prepare a seven-transparency presentation on your goals and aspirations over the next 5 years. Choose a title and material appropriate to you and to your real or imagined audience.

6.4 DELIVERY

One of the primary advantages of presenting over giving a traditional speech is the relative ease of delivery. Presenting lets you work with transparencies or a laptop, either of which lets you examine a full set of notes in disguise; moreover, the setting is less formal and is thus emotionally less demanding on the speaker. Nonetheless, the ability to deliver a presentation effectively is not an inborn trait, and here we consider some key aspects of presentation delivery.

Good delivery requires attention to a number of important details:

1. Voice projection
2. Pace

3. Modulation
4. Eye contact
5. Prop manipulation

The first requirement is that you be heard. Electronic amplification is useful in a large hall, but when using a microphone, you should remember that a normal voice is sufficient to reach the farthest audience member. I once got hoarse while working with the aid of electronic amplification because I forgot I was wearing a mike. In trying to project to the back of the room, I gave myself a sore throat and my audience an earache. Don't make the same mistake; speak at a natural volume when using amplification.

Without electronic amplification, you need to project your voice to the audience members in the last row. The key to projection is to speak deliberately and to enunciate each word separately and clearly, generating a sufficient volume of air from your diaphragm. The amplitude you can achieve approaches that of shouting, but it can be sustained for a longer time.

It is also important to speak at a modest pace. I learned much about the pace of oral presentation in my 6 years at the University of Alabama. The Southern oral tradition is widely admired, but until you've settled in to listen to some good ol' tales told in the traditional manner, you can't appreciate how the unhurried pace of the storyteller contributes to the success of the tale. This lesson transfers to average presenting. In general, the presenter is in too much of a rush to get through the material. The audience will understand more when the presenter's pace is in step with the audience's listening speed. Sustaining the vowels in key words seems to help, and liquid consonants (l, m, n, ng, and v) at word endings can be held somewhat longer to create emphasis or dramatic effect.

In spoken English, neither monotones nor singsong voices are much admired. A skilled presenter modulates both the pitch and amplitude of his or her voice to help make the talk easier to listen to; at the same time modulation helps accentuate or emphasize the most important material.

Eye contact is important in building a bridge to your audience, but eye contact can pose difficulties for the novice because the nonverbal messages received from the audience can cause the presenter to forget what he or she was going to say. Compared to speech making, presenting is fairly forgiving in this regard; you are busy with your props, and the transparencies remind you of the points you are making should you be distracted by eye contact. As you gain confidence in your presenting, you will be freer to look around the room at your audience members; doing so will give you important feedback about their level of understanding and contentment. Eye contact also gives them a further opportunity to judge your feeling and sincerity. If actual eye contact still causes trouble, you can fake it by looking in between audience members. In this way, you give the appearance of making eye contact without the interruption of your thought processes. Over time, you will gain confidence in your speaking and material to the point where real eye contact will come naturally.

Learning to work with transparencies, projectors, pens, and pointers is not difficult, but the use of such props creates opportunities for the development of annoying habits. To avoid these, it is best to remember a number of do's and don'ts of prop usage:

- Don't block the projected image with your body. A good setup can help: Place the projector relatively high, or project the image onto an elevated screen.
- Don't fidget with transparencies or laptops while you speak.
- Don't use notecards with presentations. One of the nice things about presentations is that they provide cues to you and information to your audience. Trying to manipulate both notes and a presentation gracefully is next to impossible, so put all important cues on your slides and eliminate the need for additional notes.
- Avoid multiple sources of presentation material. Stick with one medium, if possible. With modern computer graphics, you can capture your PowerPoint, your video and audio clips, and other material in the presentation file. This is the way to go. If you must switch back and forth between different devices, order the material to minimize such switching.
- To point at an image, use a laser pointer or mouse-controlled pointer. Should you choose to point at the screen itself, be sure not to block the image.

This may sound like a lot to keep in mind, but good prop usage is mainly a matter of the common sense that comes from trying to visualize your presentation from the standpoint of a typical audience member.

Having reviewed all this, the only way to get better at presenting is to present.

Exploration Exercise

Deliver a presentation on your goals and aspirations over the next 5 years, taking no more than 10 minutes from start to finish.

SUMMARY

This chapter has examined key aspects of making effective business presentations. The key notion is to present—to give a talk supported by visual aids—not speak. Busy businesspeople make a presentation more often than they give a speech, and you should do likewise. We've also considered the importance of audience analysis and subject selection and have enumerated some of the important elements of a typical presentation.

This has led to a consideration of the process of presentation development itself, together with a more detailed examination of some of the aspects of transparency preparation, including copy writing and PowerPoint slide production. Finally, we've considered key aspects of presentation delivery, including the importance of vocal projection, pace, and modulation. While there is a lot to master, by choosing to present (not speechify) you are already well on your way to becoming effective on your feet in front of a business audience.

EXERCISES

1. Prepare and deliver a brief presentation using a current topic involving engineering, engineers, or technology.

2. Prepare and deliver a brief autobiographical presentation.

3. Prepare and deliver a brief biographical presentation on an interesting family member or an acquaintance.

4. Prepare and deliver a brief sales presentation for a product or service with which you are familiar.

5. Prepare and deliver a brief sales presentation that pitches your potential as an employee to a potential employer.

6. Prepare and deliver a technically accurate presentation (for a lay audience) on a technical topic of your choosing.

7. Prepare and deliver a technical presentation (for an engineering audience) on a topic of your choosing.

8. Prepare and deliver a brief how-to presentation on some aspect of a favorite sport or hobby.

9. Form a team of five members and choose one of the topics of the previous exercises. In a brief planning session, divide the topic into subtopics for a group presentation. Prepare and deliver the presentation.

10. Have an audience select a topic of common knowledge. After taking a 10-minute period to organize your thoughts and prepare a short presentation file or handwritten transparencies, deliver a brief presentation on the selected topic.

Chapter 7

Human Side of Engineering

7.1 HUMAN CHALLENGES OF ENGINEERING

There is a fairly sharp discontinuity between schooling and working in an engineer's career. Engineering school can be a solitary affair. Long hours doing problem sets with nothing but a calculator, a pad of green engineer's paper, and a 0.5-mm mechanical pencil may be good preparation for engineering's technical challenges, but the isolation that results is not good preparation for the human side of engineering. Team and capstone design projects explore some of the human challenges of the entrepreneurial engineer's world, but the meetings, the phone calls, the client contact, and the time spent with co-workers can all add up to a level of human interaction well beyond what an engineering education prepares an engineer for.

I remember being surprised by the human relations challenges of my first full-time job, and I did what any red-blooded boy born with a library card in his hand would do: I read. I read about human relations, about sales, about marketing, about organizational behavior, and about leadership. And I made mistakes.

Not little mistakes. Mistakes that lost me friends, mistakes that lost my company sales, mistakes that ultimately sent me packing back to graduate school. And if you're wondering whether someone with such a lousy track record should be writing a chapter on human relations (or more importantly, whether you should be listening to him), so am I. But my reading and my mistakes have led to a somewhat better batting average in this ballpark, and maybe it's better to learn from the .190 hitter who has raised his average to the mid-.200s than from the batter who has always swung a .310 bat. The self-made batter knows something about improvement, while the natural has long forgotten—if he ever knew—how he came to be so good.

To start a useful dialog on human relations we must forget about your favorite person—you—and try to look at life through the eyes of others. This is the one axiom of human relations, but a number of theorems and corollaries follow directly from it. In particular, we must consider the role of praise and criticism as well as the importance of asking questions in dealing with others.

The Entrepreneurial Engineer, by David E. Goldberg
Copyright © 2006 John Wiley & Sons, Inc.

7.2 THROUGH THE EYES OF OTHERS

Human behavior is extraordinarily complex, and attempts to simplify the topic risk being naive, ineffective, or both. On the other hand, success as an engineer depends upon a practical ability to figure out the people around us, to work on teams in organizations, and to deal with clients and others outside our organizations. This requires a straightforward approach to understanding and working with people, an approach that does not require a Ph.D. in psychology or an advanced degree in organizational behavior. Yet, with so many different individuals, with their myriad motivations, multitude of life experiences, and variety of temperaments, it seems that any attempt to develop a straightforward approach to human relations would be doomed from the start. What, after all, is common among the individuals we meet in business?

Remarkably, there is one way in which we are all alike, and although it is the source of much conflict between individuals, ironically it is also that which permits us to predict the behavior of others:

We are all self-interested.

Let's face it. You are more interested in yourself than am I interested in you; I am more interested in myself than are you are interested in me. And this holds reasonably true across most pairings of individuals you can name. Of course, there is nothing strange in all this. Biologically, we are organisms programmed to survive, to look out for our own care and feeding above almost all else. As thinking beings, we also devote large proportions of our very large brains to thinking about our higher-level yearnings, wants, and needs. Such natural self-interest sounds counterproductive when it comes to building good group relations, but in a strange way it is the starting point, because if we know that we are self-interested and we know that others are self-interested, we can often predict their behavior.

Thus the starting point of good human relations is seeing things through the eyes of others. If we can understand what makes someone tick, we can start to predict what they might do under a particular set of circumstances. And once we have the ability to model effectively, as good engineers we know we have the ability to design. To build functioning circuits, we must have a model of how components react and how circuits perform. To build functioning relationships, we must have a model of how individuals behave and interact. Of course, as engineers we use both formal and informal models all the time, but the kind of modeling suggested here is somewhat more inexact and nondeterministic. Comparing our ability to anticipate human behavior to our ability to analyze a circuit, we realize that the human modeling contains more surprises, more randomness and caprice, and much less precision. On the other hand, that our modeling of human behavior is in some ways less reliable than the engineering kind does not mean it is less useful; perhaps the remarkable thing is that a single straightforward principle results in pretty good ballpark predictions of how people react.

Although the principle of seeing things through the eyes of others is straightforward, its application requires considerable skill. In the remainder of this chapter we consider its use in conflicts, conversations, and persuasive situations. We also consider the important roles of praise, criticism, and apology in our dealings with others.

7.3   ANATOMY OF A DISAGREEMENT

Nowhere is the importance of seeing things through the eyes of others more evident than in analyzing the typical disagreement. Sometimes disagreements occur because two (or more) individuals truly have an irreconcilable difference of opinion, but more often than not, one or more of the parties has not taken the time to view the situation from the opposing side.

Consider a case in point. A new engineer in a consulting company was assigned to a project by his boss. The job required that the engineer manage a group of technicians in the mapping of the piping of a chemical plant and the construction of a computer database for the project. The engineer had just completed a graduate degree and was expecting to do work that was more technically challenging than this. His negative attitude toward the assignment came across in his first on-site meeting with the client at the chemical plant when a disagreement arose over the scope of the contract. In trying to get the work to be more technically "interesting," the engineer tried to change the scope of work. This in turn disturbed the client, who complained to the engineer's boss that the work be performed as contracted. Upon returning to headquarters, the engineer got into quite a row with his boss. On the one hand, the boss could not understand how a new employee could be so cavalier with an important client and so bold as to try to change the terms of a contract without authorization. On the other hand, the engineer could not understand how he could be asked to do what he viewed as mere grunt work; this was not the kind of assignment that had been discussed when he interviewed with the company.

I wish I could report that the boss and the engineer resolved their differences and lived happily ever after, but, unfortunately, getting off on the wrong foot set the tone for the engineer's short tenure with this firm. Within 6 months he left to take another job. It didn't have to happen—if he or his boss had recognized their differences in perception, perhaps the problem could have been worked out. Without a willingness to see the conflict through the other person's eyes, however, there was little chance to reconcile these disparate views.

I should point out that this particular situation arose largely because a person fresh out of school did not consider run-of-the-mill engineering work to be "technical enough." This is a common complaint among engineers and reflects a particular pair of differing perceptions. In school, engineers are exposed to all kinds of fancy technical tools, but in practice the job that needs doing often doesn't draw on that technical tool kit. Thus the engineer is often guilty of not seeing his job through his employer's eyes, eyes that focus on the primacy of the job and the importance of getting it done. Of course, most employers are guilty of not understanding this mismatch between their engineers' expectations and the realities that they face. The engineer's usual sink-or-swim training does little to smooth the road between school and work. This book is an attempt to smooth that road from the new engineer's point of view, but it would be useful as well if employers better understood their engineers' initial orientation and tried to get them to better understand what is required of them.

Beyond these specialized conflicts between new engineers and their employers, it is true that in many arguments, people take firm positions, viewing their side as right and the other side as wrong. There is rarely black or white in human affairs; there are more often shades of gray. Moreover, even when one side is largely right, there is generally no court of appeals to declare a winner and decide how to proceed. In the garden-variety disagreement, if the arguing parties don't work through their difficulty, the knowledge of being in the right can be little comfort in smoothing the ill consequences of the impasse. And in most cases, if there is fault, there is fault enough to go around, so much so that it is useless to try to assess blame. It is better if people try to understand each other's point of view, to separate fact from perception, and to work out a practical way to proceed.

At this point we've done little to solve such misunderstandings. In a moment we'll discuss the handiest habit for encouraging the communication that can bring about greater understanding and fewer disagreements, but before we do that, it might be useful to analyze one of your own recent conflicts from the opposing side.

Exploration Exercise

Consider a recent disagreement in which you were directly involved. Write a paragraph or two analyzing the problem from the other party's point of view. Then consider ways in which the disagreement might have been avoided.

Conflict is one way in which mismatches in perception manifest themselves. In the next section, we consider how the professional salesperson or persuader can pay close attention to perceptions, thereby minimizing conflict and maximizing agreement in sales and other situations where persuasive skills must be applied.

7.4 WE ARE ALL SALESPEOPLE ON THIS BUS

Mention the word *salesperson* to an engineer and you may not get a pleasant reaction. Whether the stereotype we hold comes from the Arthur Miller play *Death of a Salesman* or from our own bad encounters with Willy Loman glad-handers having the gift for gab—and deception—the stereotype does us a disservice because it prevents us from appreciating and identifying good sales technique.

But who cares? In a book aimed at discussing life skills for engineers, why should we care about good sales technique? Sure, a few among us will earn their pay by pounding the streets of technical sales, but those people receive separate training. Why take time out of our busy agenda to stop and talk about powers of persuasion? The answer to all of these questions is a single word: ideas. As engineers, our primary stock in trade is ideas, oftentimes innovative ideas, ideas that have not been tried, ideas that may encounter stiff resistance from co-workers, bosses, clients, or consumers. These people and others must be persuaded—they must be sold—before they are willing to give a new idea a try. Nothing can be quite as frustrating to an engineer as to have a good idea but to be unable to get anyone to take a look at it. This situation is further exacerbated by engineers who often find the logic of their own arguments compelling, so compelling they feel that the world should beat a path to their door. Once again we are confronted by a perceptual problem. The world—our co-workers, bosses, clients, or consumers—often does not see things the way we do—may not think the way we do. If we are to be successful in gaining the acceptance of our ideas—if we are to be successful engineers—we must try to narrow the perceptual divide between us and those we seek to persuade.

Having established that, in a sense, we engineers must all be salespeople at one time or another, we can now ask, what makes a good one? When I worked in the engineering-software business, I hit the streets looking for business, and over the years I had the opportunity to get to know a number of professional salespeople. I found it interesting how far out of line the Willy Loman stereotype was with the behavior of these sales pros. Indeed, the best ones were confident and fairly fearless, but those characteristics were not necessarily the ones that started them on the road to sales success. Usually what separated the stars from the meteorites was their ability to listen. Perhaps this flies in the face of conventional wisdom, but a salesperson cannot force you to buy something you don't want. The only real option he or she has is to show you how some product fulfills some physical or psychological need you have. And the only way to find out what that need is is to probe and to listen.

We will consider ways to enhance our listening capability in a moment, but here it is interesting to consider that the stereotypical salesperson—the one who does all the talking—is the antithesis of the effective listener-persuader. The reason we remember the gift-of-gab guys and gals is that they annoyed us so. (Inevitably we walked out before buying anything, or if we did buy, we almost immediately regretted it.) On the other hand, it is easy to forget the good salespeople. They're so smooth we often think that they just wrote up our order, but careful analysis of many such situations reveals an effective listener, matching need with product to facilitate an easier decision.

Let's see if you can extract a human relations lesson in persuasion from your own recent encounters with salespeople in the next exploration exercise.

Exploration Exercise

Consider two recent experiences with salespeople, one good and one bad. In each case, consider how much they talked versus how much they listened. Compare and contrast how much each salesperson thought in terms of his or her needs versus yours.

Thus far we have considered situations of conflict and situations of persuasion. In both cases, individuals are seeking a change in the status quo, and in both cases a perceptual gap exists. Conflicts arise from a lack of attention to other's perceptions, and sales occur with devoted attention to others' views. The vital question is how does one become more adept at seeing things from another person's perspective? In the next section, we consider the crucial role questions play in this regard.

7.5 THE ROLE OF QUESTIONS

The key to human relations is seeing things through the eyes of others, and the key to seeing things through the eyes of others is *asking questions*. Once this is said it is easy enough to understand, but it is surprising that so many people believe that the way to resolve a conflict, make a sale, initiate a friendship, be a good conversationalist, or conduct just about any activity involving others is to tell their side of the story. This approach discounts the interests of the other person, who is as egotistical and self-centered as we are and who will be appeased, persuaded, friendly, conversational, or in other ways more positive toward us if given the chance to express his or her views.

The most effective way to draw people out is to ask questions. In this section, we consider the asking of questions in differing circumstances, including conversational, conflict-resolving, and persuasive situations. We will see that different kinds of questions are appropriate in different situations and will identify some of the more important types.

7.5.1 Questions in Conversation

"I'm not a very good conversationalist. I never know what to say." How many times have you heard someone say something similar to that? No doubt those same people have had good conversations, but it is difficult to stand back from our own human interactions and understand what has transpired. What characterizes good conversation? Usually in good conversation, at least one of the parties asks a question and then listens carefully to the response, following up with more questions that move the conversation along. Thus the prime mover of a conversation is not the talker—that is the easy role. The motive force behind every conversation is the questioner-listener. Of course, the best conversations are those where the questioning and talking roles are exchanged repeatedly.

What types of questions can move a conversation along? It is difficult to generalize, but open-ended questions about something that interests a conversation partner aren't a bad place to start. After all, we know that other people are most interested in themselves.

7.5.2 Questions in Conflict Resolution and Negotiation

Conflict resolution also requires the use of questions but from a more elaborate approach. Whether dealing with an interpersonal problem or an organization-wide conflict, it is important to use various questioning techniques to narrow the perceptual gap that exists between the parties involved. Such a conflict resolution episode typically begins with the recognition by one or more parties that a problem exists. Once this occurs, one of the parties must observe, "We've got a problem. How do you see it?" or something similar. This fairly open-ended approach invites the other party to share his or her frustrations. When that person finishes, the questioner can briefly summarize what he or she has heard and ask whether the summarized view is valid. If it's not, a more directed question or two can iron out differences, and, within a few iterations, the process should achieve perceptual convergence. At this point, the original questioner might ask whether it would be all right to share his or her view of the problem. The original talker is now the listener, and after the other view is shared, the listener is asked to summarize what he or she has heard.

After perceptual convergence on this second party's view, a series of questions can then be asked to identify the differences between the two individuals. This series can be followed by a series of more specific questions to see whether there is any room for maneuver or compromise. The progression from open-ended, to confirmatory, to increasingly directed questions moves the parties from conflict toward points of agreement and possible compromise, and closer to the resolution of the conflict. If the conflict cannot be resolved, at least the parties will know that it was not for lack of understanding but rather because of truly irreconcilable differences.

Exploration Exercise

Consider a recent disagreement in which you were directly involved. Make a list of 10 questions you might have asked the other party to probe his or her position and perception. Make a list of 10 questions he or she might have asked you.

7.5.3 Questions in Sales and Persuasion

Persuasive situations call for all the questioning skill a persuader can muster. We will consider a formal sales cycle as our model situation. In a professional sales situation, questioning usually begins along conversational lines in an attempt to probe the customer's interests, motivations, and needs in a general sense.

After identifying needs connected with the product or service, the salesperson may begin a perception-confirming sequence of questions and summaries, and confirmatory-type questions may begin, though there is usually no need for the persuader to share his or her perceptions with the customer.

After establishing a few perceptual outposts, the persuader can narrow the questioning to more specific lines, that is, to what a professional salesperson calls *closing* questions. The bottom line in all persuasion situations is that the business must be asked for and gotten. Books on sales are a better place to read about this well-developed art form, but some of the more salient types of "closes" can be discussed here.

The conditional close is a good initial trial balloon, and it goes something like this: "If I can show you that X, Y, and Z occur as a result of using this product, will you buy?" If the person says yes, it is then a matter of persuading him or her that X, Y, and Z will occur. If the person says no, there is then an opportunity to ask what conditions still obstruct the sale. Along the way, this kind of questioning can lead to the discovery of one or more such objections; the uncovering of objections is a call to return to a more open-ended form of questioning to obtain perceptual convergence on the customer's buying blocks, thereby paving the way for their removal.

After removing objections, further closing attempts can be made from the direct close, "Would you like to buy this today?" to the somewhat sneaky assumptive close, "Would you like it in red or in blue?" In this way, the persuader can work from open-ended, information-gathering questions, to more specific needs-defining questions, to the closing questions that clinch the deal.

Skill in the art of the question can help make us surprisingly effective in our dealings with people. In our increasingly electronic, anonymous society, the art of conversation and one-on-one communication is being lost. It can be regained if we only take time to ask.

7.6 PRAISE

Beyond the desire to be understood and listened to, each of us loves to be praised. Children adore the praise of their parents. Spouses crave the praise of each other. Workers crave the praise of bosses and co-workers. Despite our ravenous appetite for praise, we are remarkably stingy in handing it out; of course, this represents a remarkable breakdown in our seeing things through the eyes of others (Carnegie, 1981).

Why are we so tight with our praise? Do we see it as a kind of currency to be hoarded? Do we view this praising business as some sort of zero-sum game with only so much to go around, so that the praising of others may lead to their success at the expense of our own? Such fears are rarely warranted. Far from being inherently scarce, praise is a fully renewable resource, with many people around us doing praiseworthy things and only ourselves to blame if we don't make the time or effort to notice them.

And it's unfortunate that we don't take more time to notice because praise works a powerful magic on the people it touches. I recall remarking to a frowning, somewhat grumpy woman behind a rental-car counter at Washington National Airport what nice handwriting she had. A large smile came over her face and we had a nice chat about business and the weather. After this nice chat, and without my asking, she took special care to give me a brand-new car with only 23 miles on it. Understand, I told her that she had good handwriting not because I wanted a new car or anything else from her. I praised her handwriting because she indeed had lovely handwriting, and what looked as though it might become a stereotypically bad service experience turned into a pleasant human encounter.

Of course, there may be times when you do praise with the hope of improvement or change. A young engineer joined a major consulting firm and noticed that the janitorial service in her office was spotty at best. She noticed that the same janitor worked in the department each day, and so one day when the janitor had done a better-than-usual job, she stopped him in the hall and said that she appreciated the extra effort put into the cleaning that day and that she really appreciated it when he took special care in sweeping and dusting her room. The janitor seemed stunned that someone had noticed the extra effort and said something about thinking that no one cared about cleanliness these days. The engineer assured him that she certainly did and that she was glad to have someone working in her part of the building with an old-fashioned attitude toward neatness. Shortly after this conversation the janitor instituted a spot-waxing program in the building, enlisting the help of the other building janitors, and until that janitor was transferred to another building, the engineer never noticed another lapse in cleanliness.

While we should recognize that praise is something we all crave and that it can have a remarkable effect on people, we should guard against that imposter, *flattery*. Flattery resembles praise in that it compliments a person for something, but it lacks sincerity. Individuals who have inflated opinions of themselves can be flattered (i.e., all of us can be flattered at least some of the time), but in better moments most of us can recognize flattery as the imposter it is. When recognized, flattery can have a worse effect than never having said anything at all.

To distinguish heartfelt praise from flattery, it is helpful to be truthful and specific when praising. When you say something nice, say exactly what it is you like. For example, in my encounter at the airport, I did not say something vague about the woman's appearance or demeanor; I said I thought she had nice handwriting, and she did. If you say something specific, there will be less chance that your comment will be misinterpreted as mere flattery.

Exploration Exercise

Consider a person with whom you have regular contact. Make a list of several things you truly like about that person. On your next meeting, at an appropriate time, offer praise regarding one of the things you most like. Write a brief paragraph reporting what you praised, why, and the individual's reaction.

7.7 CRITICISM

If it is important to recognize those around us for the things they do that we like, it is equally important to be cautious in criticizing them when they do things we don't like; when criticism is necessary, it is important to express it in a way that targets the behavior and not the person.

That criticism should be used cautiously is not surprising advice if only we see things through the eyes of others. When was the last time you enjoyed being criticized? I realize, in retrospect, that there were times when I needed to be criticized, but I can't recall a single instance when I was happy or particularly grateful for it at the time. I remember many times when I felt that criticism was overly harsh or disproportionate to the crime, and I can think of a number of people I am less than fond of, largely as a result of their critical nature. We all have similar feelings, and the projection of our feelings onto others should be fairly immediate: If we don't like receiving criticism, why should anyone else?

The other thing wrong with criticism is that it can easily be ineffective. Many people have self-defense mechanisms with an enormous capacity to deflect criticism. If armed robbers, rapists, child molesters, and even cold-blooded killers can rationalize their savage, immoral behavior, the average Joe or Jane can certainly deflect accusations of petty wrongdoing. If we are interested in being effective—if we are interested in changing behavior—we must maintain the confidence and trust of the people who have done wrong and help them see why it is to their benefit to change their ways.

There are many ways to accomplish this. One is to offer criticism in a spirit of helpfulness. This is a fairly direct approach, and its directness occasionally can lead to resentment; however, words such as "I know you are giving your best effort, and do you think it would be possible if you tried XYZ?" can sometimes temper the blow enough to make a breakthrough. Notice that phrasing the constructive criticism in the form of a question has the effect of tempering the blow even further. Also notice how the use of "and" rather than "but" helps prevent the erection of additional psychological barriers before the sentence is finished.

Another way to temper criticism is to point to your own failings. Sometimes telling a little story about a personal mistake before you ask a person to change his or her behavior is an effective means of offering criticism. It can also be helpful to play down the mistake the person has made. If you make the mistake sound like a big deal, requiring a big effort to rectify, the artificially high hurdle you've erected will make the person resist changing all the more. If you make the error seem easy to correct, you should encounter less resistance to your suggestions.

In addition, it is important not to spend time assessing blame. Some time ago, I had a boss who spent a good portion of his day tracking down mistakes and those who made them. That attitude paralyzed the whole organization, to the point that no one did very much for fear of making a mistake. The proper approach to mistakes is that, once they are uncovered, they be corrected quickly and the individuals try harder to avoid them. Looking to assess blame only makes people more secretive and less cooperative in fixing problems.

It is useful to turn this reasoning on its head at times and use error count as a productivity indicator. The reasoning goes something like this. Given that we try hard to improve on our mistakes, our rate of error tends to remain constant or diminish over time. Therefore the number of errors we make is at least proportional to the amount of work we are doing. We would always rather that errors not occur; but given that they do, and always will, occur, seeing something positive in their occurrence can help us adopt an attitude that contributes to their correction and reduction.

Exploration Exercise

Analyze a recent situation in which you were criticized or in which you criticized someone. In two paragraphs describe the situation as it occurred and then describe how it might have been handled otherwise.

7.8 ENGINEERING IS SOMETIMES HAVING TO SAY YOU'RE SORRY

On the subject of mistakes, it seems only fair to consider our own. At the same time we are lightning quick to point out the errors of others, we can be glacially slow to admit our own.

To recognize the rarity of apologies, we need only ask and answer two questions:

1. When was the last time someone apologized to you?
2. When was the last time you apologized to someone?

I don't know your answers to these questions, but for many readers it has probably been a long time since they have made or received an apology. What is it that makes us so unable to admit our mistakes, apologize, and move on? Perhaps it is a matter of excessive pride combined with a fear of appearing weak. There is little to do about excessive pride but try to overcome it. On the other hand, apologizing when you have made a mistake, far from projecting weakness, will often be seen as a sign of strength of character and confidence.

One of the reasons we may not want to apologize is that we may feel we are only partially at fault; we think that if we apologize the other person will blame us entirely without examining his or her role at all. In situations such as this, a good approach is the conditional, or explained, apology. In an explained apology you begin by calmly and briefly explaining what irritated you about the other person's behavior, but you go on to say that, regardless of anyone else's behavior, you have no excuse for behaving as badly as you did and you apologize. Forced to face having had a role in the problem, many will admit their contribution, and the relationship can proceed with little damage. Others will not see their role, or will refuse to admit it. In these cases, the person making the explained apology can determine whether there is value in using questions to try to get the other person to recognize his or her role in the conflict. In either case, the explained apology can often clear the air sufficiently to allow the business at hand to proceed.

Exploration Exercise

Consider one recent incident where you felt a person should have apologized to you; then consider one recent incident where you feel you should have apologized to someone else. In each case, consider whether your relationship with each of the individuals has become worse, better, or remained the same. Write a paragraph describing each incident and the change in relationship that occurred thereafter.

7.9 WEAR A LITTLE PASSION

In many situations, it is possible to "view the glass as being half empty or half full." As engineers, our passion for logic tells us that the choice should be a matter of some indifference. As human beings, our passion for passion suggests that we should choose to view the full portion because we know that our perceptions of situations can be profoundly affected by our attitudes toward them. Achieving success as an engineer requires persistent pursuit of intermediate and long-term goals over an extended period of time, and such persistence is much easier to sustain if we approach our work—and our lives—with zest and enthusiasm.

People show their passions in different ways, and I am not recommending that we all wear our emotions on our sleeves. I am suggesting, however, that some externalization of our positive emotions can have a positive impact on our approach to the challenges of life, and can also help brighten the world of those around us. Organizations that learn to wear a little passion have a positive glow of productivity about them. Of course, it is just as easy, perhaps easier, for organizations and the individuals within them to become gripped by a negative, can't-do attitude. Such working environments are not a pretty sight.

There are a number of practical ways to stay positively oriented; one of the most important is to be doing something you enjoy—something you find engaging. Another habit is to simply smile and laugh more often. Smiles and laughter are contagious and can have therapeutic value when things aren't going just right.

Another helpful habit is trying to emphasize the good that often eventually comes from initially stressful situations. Many clouds *do* have a silver lining, and we would all do well to spend more time thinking about the eventual positive consequences of today's mishaps. Moreover, we should make efforts to see the current good in bad situations. The doughnut may presently have too large a hole, but there is still an edible portion.

Together these habits can help us get past everyday stresses and obstacles and reach eventual success.

SUMMARY

This chapter has examined the importance of good human relations and has considered a number of ways to help us smooth our dealings with other people. The basic principle from which all good human relations flow is seeing things through the eyes of others. Asking questions helps us make the critical shift from our own viewpoint to that of another person. If questions help get us into their minds, praise helps get us into their hearts, helping them feel good about the things they do well.

We have also considered the harm that can come from criticism and have suggested that criticism should be applied cautiously and with care. We have considered how apologies should be offered when we discover one of our own mistakes. We have examined some of the reasons why apologizing is so difficult. We have even looked at a way to apologize when another person may also bear some of the responsibility for a conflict. The role of enthusiasm in keeping ourselves and those around us upbeat, positive, and looking forward to the challenges ahead has also been considered.

EXERCISES

1. During the course of a day, make a list of your mistakes, both big and small. Write a short paragraph considering whether the number is larger or smaller than you thought it would be.

2. Keep two lists during the course of a week. On one, record the number of times you are complimented and on the other record the number of times you are criticized. Write two paragraphs comparing and contrasting the quantity, quality, and severity of praise and criticism you experience.

3. During the course of a day, couch all requests for action in the form of questions (e.g., "Could you do X?"). During the course of the next day, give all orders as commands (e.g., "Do X."). Write a paragraph comparing and contrasting the response to the two approaches.

4. Imagine that you are being interviewed for a job. Make a list of 10 questions your interviewer might ask. Make a list of 10 questions to help clarify, deflect, or redirect an interviewer's questions when they are unclear, unanswerable, or inappropriate.

5. Imagine you are a company representative sent to interview a candidate for a job. Write a paragraph describing specific characteristics of the ideal candidate. Now think of these characteristics from the interviewer's perspective. What ramifications does each have for a potential job candidate's interviewing behavior?

6. At a social occasion, make an effort to hold two types of conversations. In one, make statements and assertions. In the other, ask lots of questions. Write two paragraphs comparing and contrasting the two types of conversation.

7. Write a brief essay describing the characteristics of an employee your boss would want to have. Discuss the ramifications for your own behavior.

8. An engineer in your group has told you that an engineer on another team has presented an idea of yours as his own. In a short paragraph describe the steps, if any, you would take to handle such a situation.

9. Bill, an engineer in your firm, has written you a "flame," a highly critical e-mail message chastising you for some design work you did a year ago. Should you fire back a return flame to Bill, call him on the phone, see him in person, write a critical

memo to his boss, spread rumors questioning his mental state, write a reply via official memorandum, or take some other action? In a brief paragraph describe why you selected your particular course of action. Also, compare and contrast the effectiveness of in-person visits, e-mail, written memos, and the telephone in handling negative situations.

10. Joan, an engineer in your firm, has performed superbly in connection with a project you're working on. Should you ignore her, take credit for her actions, thank her in person, thank her by phone, thank her by letter with copies to appropriate managers, or take some other action? In a brief paragraph describe why you selected your particular course of action. Also, compare and contrast the effectiveness of in-person visits, e-mail, written memos, and the telephone in handling positive situations.

11. Form a team of three and role play a hypothetical job interview, with two of the group playing interviewers and the third playing the job candidate. Take turns until everyone has played the candidate.

12. Form a group of three and role play a hypothetical sales call, with one group member acting as the salesperson and the other two acting as prospective buyers at the same target company. Exchange roles until everyone has taken a turn as the salesperson.

13. Form a group of five. Sit in a circle and take turns offering praise to other members of the group. After each statement of praise, the other group members may challenge the remark by saying, "Flattery." If two or more members say, "Flattery," the praiser gets no points. If one or none says, "Flattery," the praiser receives a point. After five rounds, the praiser with the most points wins.

Chapter 8

Ethics in Matters Small, Large, and Engineering

8.1 IS ENGINEERING ETHICS NECESSARILY A DREADFUL BORE?

Engineering ethics is a topic that can cause an engineer's eyes to glaze over, and for good reason. What I remember of the few ethics lectures I was forced to attend in engineering school consists of platitudes, wordy, abstract codes, and hypothetical cases concocted to stimulate our thinking. Unfortunately, my fellow engineering students and I weren't very stimulated back then, and the lack of success of the usual approach to engineering or business ethics is now reflected in headline after headline decrying one ethical or moral failure after another. Student cheating is up, CEO fiduciary responsibility is down, and major technology failures occur because engineers stretch the technical truth with modest amounts of pressure from some managerial higher up.

Of course, there are no guarantees that *any* approach to teaching and learning engineering or ethics will fix a particular set of ethical lapses, but the view taken here is that the probability of taking the lessons of engineering ethics to heart increase in proportion to how down to earth the study is. This volume has been a down-to-earth affair, and our approach to ethics will stay that course.

One way to combat ethics fatigue is to start with familiar matters, so we start our investigation of ethics by considering two flavors of golden rule: positive and negative golden rules. Golden rules ask us to behave consistently or charitably depending on the flavor, but other types of ethical reasoning have been studied over the years and a number of these must be discussed. Interestingly, from an engineering point of view, different modes of ethical thought can be viewed in almost dynamic systems terms, and we should make some effort to do so. Oftentimes, human beings have ethical mishaps, not because they didn't know the right thing to do, but because they couldn't bring themselves to do it. Our own self-interest, obedience to authority, and conformity to the group are three obstacles to doing the right thing, and we need to understand how powerful each of these influences can be.

The Entrepreneurial Engineer, by David E. Goldberg
Copyright © 2006 John Wiley & Sons, Inc.

Having said this, ethical practice on the small stuff can leave us in good stead when it comes time to do right on something big. This inductive approach, working from the small details outward, helps us move from personal ethics to engineering ethics. In tackling engineering ethics, we start by studying the notion of a *profession* and find that having a code of ethics is considered by many a *sine qua non* of professional existence. We study two very different engineering codes of ethics and consider why they might be so different. This leads us to define and sketch out the notion of a *conflict of interest*, which leads us to consider the consequences and some alternatives to whistleblowing.

8.2 ETHICS: SYSTEMATIC STUDY OF RIGHT AND WRONG

Ethics is a highfalutin word, but at root ethics is the study of right and wrong. The modern study of ethics is a convoluted one, but the twists and turns of sophisticated intellectual debate obscure a much simpler truth. Human parents since time immemorial have taught their children right conduct and have tried to teach them to avoid wrong conduct, and over much of human history increasingly larger social structures (families, tribes, congregations, communities, etc.) have been the focal point for capturing and codifying rules of right and wrong for continued cultural transmission of these important snippets of wisdom. Interestingly the details of what has been considered right and wrong have varied in time and space, but these early practitioners of ethical behavior were largely in agreement about the key ideas, and most of the planet's early and great cultures arrived at one of the forms of a *golden rule*.

8.2.1 Golden Rules: Positive and Negative

Many of the great religions and cultures of the world have arrived at some form of the golden rule (Table 8.1).

Search the Web for the term "golden rule" and you'll quickly find many more. Although the wordings are different, the meanings are virtually the same. Note, however, there are two different types of golden rule (Hazlitt, 1964).

Negative Golden Rules

Negative golden rules are in many ways the more practical of the two as their dictates are easier to fulfill. Consider, for example, the Confucian golden rule, a negative golden rule:

Do not do to others what you would not like yourself.

The rule urges us to consider what we would *not* like to have done to us and then *not* do those things to others. Thus, the rule seeks to *inhibit* behavior that is likely to be hurtful to others. Note that the default action under a negative golden rule is to avoid acts toward others: to leave others alone. This makes common

Table 8.1 Some of the World's Great Golden Rules

Religion	Golden Rule
Buddhism	Hurt not others in ways that you yourself would find hurtful. (Udana-Varga 5:18)
Christianity	All things whatsoever ye would that men should do to you, do ye so to them; for this is the law and the prophets. (Matthew 7:12)
Confucianism	Do not do to others what you would not like yourself. Then there will be no resentment against you, either in the family or in the state. (Analects 12:2)
Hinduism	This is the sum of the Dharma [duty]: do naught onto others what would cause pain if done unto you. (Mahabharata 5:1517)
Islam	None of you truly believes until you wish for your brother what you wish for yourself. (Sunnah)
Judaism	Thou shalt love thy neighbor as thyself. (Leviticus 19:18) What is hateful to you, do not do to your neighbor. This is the whole Torah; all the rest is commentary. (Talmud, Shabbat 31a)

sense as most of us have just wondered why others don't just "leave us alone," and negative golden rules encourage us to that end.

While negative rules are useful rough-and-ready guides to avoiding bad conduct, they are less helpful in a philosophically rigorous analysis. The primary criticism of negative golden rules is that they leave the determination of wrong conduct in the hands of a single individual: you. That is, each individual is to judge whether or not an action is improper by his or her own belief system. In a homogeneous population with a widely shared set of beliefs, the widespread following of a negative golden rule can result in the diminution or absence of large-scale social disharmony. Where beliefs vary a good bit—in culturally or ethnically diverse populations, for example—following a negative golden rule may not be sufficient to avoid conflict. Individuals will differ too much on what constitutes right and wrong.

Positive Golden Rules

Positive rules go a step beyond negative golden rules, requiring *right conduct,* even *exemplary,* conduct toward all. Consider, for example, the Christian golden rule, a positive golden rule:

> All things whatsoever ye would that men would do to you, do ye so to them.

The rule urges us to consider what we would like to have done to us and then *do* those things to others. Since positive golden rules focus on *doing* versus *not doing*, they are more activist in nature, and the requirement to do good is limitless in magnitude and without end. For example, I would very much like my publisher to increase my royalty for writing these words to $1,000,000 (in small unmarked bills) per book. Does my unreasonable desire require that I myself should offer to pay my barber a similarly outrageous sum for my next haircut? The infinite ways in which we can be generous to our fellow human beings make it certain that we will not do so at every possible opportunity.

Clearly, then, taken to their limits, positive golden rules are impractical on their face, but this misses their point. The intent of positive golden rules is to set up an *ideal* for good or even righteous conduct toward our fellow human beings. Positive golden rules hold us to a higher standard than negative golden rules, asking us not only to avoid doing harm toward others, but to do good deeds widely and often. In other words, they ask us not only to avoid doing bad things, but to treat others kindly and well.

Exploration Exercise

Consider an incident in your life where you consciously applied a golden rule. Write a short paragraph describing the incident, whether the golden rule was a positive or negative rule, and the ways in which your behavior was different than what you might have otherwise done. If you cannot think of such an incident, describe an event where you believe someone else was acting according to a golden rule.

8.2.2 Whence Right and Wrong?

The almost universal nature of golden rules makes them almost irresistible as a centerpiece of moral theory, but moral philosophers have long recognized their Achilles heel. Since golden rules—whether positive or negative—are defined in terms of individual-centered notions of right and wrong, they depend on widely shared notions of right and wrong conduct as *input*. This places a premium on knowing where individuals get a sense of right and wrong in the first place. Moral philosophy since ancient times has jumped through a variety of intellectual hoops trying to find a satisfactory answer to this question. As with other branches of philosophy, moral philosophy returns with not a single answer but a number of answers, each with its own set of merits and defects. Here we consider a number of broad categories of answer to the title question of the section as follows:

1. Right and wrong come from culturally or religiously determined norms of conduct.
2. Right and wrong come from an innate moral sense.
3. Right and wrong come from maximizing human happiness or utility.
4. Right and wrong come from consistency in reasoning.

5. Right and wrong are not sensible topics of conversation for rigorous philosophers or other human beings.

Each of these points of view represents a major strain of philosophical thought, and to discuss them at length is beyond the scope of this treatment, but we do briefly consider their main points.

From a Religious or Cultural Set of Norms

Almost all of us were introduced to right and wrong at home, and many of us had views expressed at home supplemented by that of a clergyman as part of some organized religion. Outside the home, culturally acceptable norms of behavior are taught and enforced as we go to school, work, and start our own families. As a practical matter, religious and cultural norms of right and wrong are the dominant influence in our lives. The tendency in the last century by philosophers was to question these traditional influences and even question the validity of notions of right and wrong, but culture and religion remain highly influential.

Although modern philosophy has not always been particularly friendly toward ethics or other sets of beliefs that are at least, in part, socially transmitted, one element of modern philosophy relevant to understanding the propagation of belief systems is Searle's theory of the *construction of social reality* (Searle, 1995). Searle examines the philosophical basis of *social facts*, such as chess and money. Chess is not chess because of the shape of the pieces or materials of the board, but it is chess because we agree that it is a game played on an 8 × 8 board with 16 pieces per side where each piece follows a particular set of rules of movement. Likewise, money is not money because of the paper or metal of bills or coins. It is money because people assign it value and agree to exchange valuable things for it. Chess and money are strange objects in this way, but Searle argues that they are objectively real in this special social sense (and if you doubt this, please feel free to send all the currency in your wallet—or your nicest chess set—to my university address).

In moral matters, the Searlean view might be that ethical codes are ethical codes (and effective) because we and others believe in them. Moral rules (such as golden rules) work when the subject population shares similar ideas of right and wrong and practices those ideas with regularity. Returning to the example of money, a currency is widely accepted as strong when its supply is kept in check, when counterfeit money is minimized, and when a large market or markets accept it in exchange for goods and services. Although there is no counterpart to money as a medium of ethical exchange, the myriad deals we make, promises we keep, and courtesies we extend to one another form a kind of marketplace of ethical conduct. When large numbers of people start to make promises that they don't keep or when discourtesy is the rule, our currency of moral conduct is inflated or devalued in much the same way that money becomes worth less in difficult financial times. While such reasoning certainly helps us understand some of the common failure modes for ethical reasoning, it offers few fundamental principles to help society get back to more widely shared ethical behavior and beliefs.

From a Moral Sense

Once a set of moral rules exists in society, the rules can take on a life of their own, but the adoption of a set of rules at all is itself something of a puzzle. In some ways, the key question in ethics is not that murder, mayhem, and ethical mishap occur regularly and are reported in the daily newspaper. The real issue is that they are *news*; in other words, they are fairly *rare* occurrences. Instead of being surprised by mayhem, how did it become so rare? In prehistoric times, how was it that early humans developed moral rules? Why didn't they just bop one another over the head and take one another's lives and possessions. Doing so would have been perfectly rational from a posture of pure self-interest. But over time, humans have developed to the point where they take special steps to avoid conflict and generally do not interfere with one another. Given the difficulty of evolution of cooperation, it is puzzling that human ethical reasoning has come as far as it has.

One longstanding and influential line of thought that helps explain this conundrum is so-called *ethical naturalism* that asserts that human nature is imbued with a certain tendency to do right. The 18th-century philosopher Adam Smith (Smith, 1759/1984), building on the work of David Hume, is remembered for the assertion that human beings are to some extent moved by *sympathy* for others. Smith did not assert that morals are predetermined in a mechanistic way, but he saw the need to assume the existence of a moral sense to propel and sustain ethical behavior. More recently, James Q. Wilson (1993) has revived this strain of thought and has added modern results from social science and evolutionary theory to bolster the argument.

From Maximizing Utility

Adam Smith is better remembered for laying the foundations of modern economics (Smith, 1776/1937), and it is interesting that his moral philosophy took a naturalistic turn. Given the strains of intellectual thought at the time, one might have guessed that the father of economic thought would have been more of a *utilitarian.* Utilitarianism is the idea that right conduct comes from maximizing *human happiness.* In other words, right action is that which maximizes human happiness. Jeremy Bentham articulated a key utilitarian idea when he said:

> It is the greatest good to the greatest number which is the measure of right and wrong.

And J. S. Mill followed with a stout defense of utilitarianism in his volume by the same name (Mill, 1861/1993).

To engineers, there is a quantitative logic about utilitarian thought that has great appeal, and commonplace engineering procedures are utilitarian by their nature. For example, when an environmental engineer does a cost–benefit analysis, he or she is attempting to follow a utilitarian approach to right conduct.

As with most philosophical ideas, however, utilitarianism raises as many questions as it answers:

1. What is happiness?
2. Whose happiness do we maximize?
3. Can we calculate happiness accurately?
4. Are we maximizing happiness now or later?

Each of these is briefly discussed in what follows.

The question "what is happiness?" has challenged philosophers since time immemorial. Early utilitarians went back to the hedonists of ancient Greece and argued that *sensory pleasure* is human happiness, but others have rejected such a view as too simple. Mill argued that pleasure in a larger sense could be taken synonymously with happiness and many have joined in the debate. More recently, research in human happiness by modern psychologists has coined the term *flow* to describe happiness as a kind of intricate interwoven complexity that results in long-term and profound engagement with one's life activities and partners (Csikszentmihalyi, 1990). In many ways, these views harken back to Aristotle and the Greek notion of happiness called *eudaimonia,* or literally "having a good guardian spirit."

Even if we can agree upon what we mean by happiness, another issue quickly appears. Whose happiness should we maximize: yours, mine, or all of ours? Like Bentham, most utilitarians believe in some sort of the greatest good for the greatest number, and they therefore argue for calculation of utility at a level above that of a single individual. Certainly it is easy to imagine self-interested beings maximizing their own utility, but if we are to maximize the utility of society, does this suggest that the naturalists were right, that we have some intrinsic, perhaps evolutionary, tendency to do right by others?

Moreover, the suggestion that we calculate happiness leads to a question whether human beings are even capable of doing so with any accuracy. Are we able as individuals to calculate the utility of our own prospective actions on ourselves and on others? It is difficult enough to know one's own mind, let alone to have knowledge of the motivations, needs, and objectives of others. How can we calculate accurately about the aggregate utility of some large social group on any given action even if we wanted to? A famous economic argument by Hayek (1945) suggests that markets and prices are necessary because they reveal exactly the kind of information necessary to do something akin to utilitarian computation, but that the right information is only revealed if the individual actors pursue their own self-interest in a competitive market. Individuals signal their preferences through sequences of self-interested exchanges in the marketplace, and the resulting prices are used by others for all kinds of calculations. Utilitarians need to reason about the utility of their actions in relation to how they affect others, but markets reveal the utility values of others through a series of *self-interested exchanges* in the market. Without these exchanges, there is no price mechanism and no way of revealing the value attached to different actions. Hayek's argument leads to a more individualistic view on the grounds of knowledge gain, but even if the calculus of aggregate utilitarianism were possible, we might argue that the calculation of utility act by act is simply impractical.

Some argue, as a result, that *act utilitarianism*, utility theory applied to the evaluation of individual acts is flawed and argue for so-called *rule utilitarianism*. That is, we cannot calculate right and wrong from one act to the next or from moment to moment, so instead we are left with devising sets of useful *rules* that practically help guide our choice of right conduct on average. Certainly the kinds of rules of ethics that have evolved in engineering and other professions are evaluated along these lines.

Finally, all of the discussion to this point has neglected time almost entirely. Even if we agree upon what should be maximized, who it should be maximized for, and that we maximize it in practice, the issue arises whether we perform the maximization for the present moment or in the longer run. Many simplistic objections to utilitarianism assume that the utility calculator will only consider the short run and ignore the long run, but maximizing utility into the future is itself a tricky business. Our ability to predict future events is severely limited, and the question also arises regarding the *period of optimization*. Should we maximize utility for our own generation, our children's generation, our grand-children's generation, or all generations to come. Although thinking about time helps overcome simple objections to utilitarianism based on myopia, consider-ing the ramifications of one's decision for all time complicates the utilitarian computation along many of the dimensions just discussed.

From Consistency

Immanuel Kant did not react well to the arguments of the utilitarians and believed that right and wrong were too important to be left to what amounted to the balancing of a moral checkbook. His formulation depended on a different sort of calculation—a logical calculation of consistency—to determine right conduct. Starting from the assumption of *good will* to do one's *duty*, Kant formulated his *categorical imperative* which sounds to modern ears like a golden rule on steroids:

> *I am never to act otherwise than so that I could also will that my maxim should become universal law.*

In other words, only those actions that can be universally generalized to all humanity should be undertaken.

In one sense, the generality of Kant's formulation does go beyond the local ethics of simple golden rules, but in so doing it leaves us with the possibility that we may have either insufficient guidance to right conduct (because the require-ment for universal generality is too difficult) or incorrect guidance (because we have incorrectly generalized that which is not universal). Despite these concerns, Kant's advice is important to us here because it highlights a key concept that guides common moral reasoning.

Kant's formulation asks us above all to be logically *consistent.* All of us have had the experience of someone giving an order or creating a rule and then watching that same person violate the order or break the rule him- or herself. This is so commonplace that we give it a special name—*hypocrisy*—and Kant may be regarded as the premier philosophical guardian against hypocrisy.

Consistency is generally regarded as laudable, but interesting consequences follow when we value it above all else. For example, Kantian reasoning usual leads to a rejection of lying under any circumstances, but one of Kant's earliest critics asked if it is therefore necessary to tell a murderer the truth about the location of his prey. Such discussions have led to various modifications of Kant's theories, but these modifications continue to value consistency and universality fairly highly.

Not a Sensible Topic of Conversation

Greek philosophy and philosophy of the enlightenment and the 19th century deal with philosophical subjects directly. In other words, the terms of philosophical discourse (terms such as "right" and "wrong") are assumed to have meaning, and the job of the philosopher is to seek their intellectual basis. For much of the 20th and current centuries, philosophers have turned to *metalevel* analysis of philosophical discourse itself. In doing so, a number of philosophers have found the language of moral philosophy to be an empty vessel.

Some, such as Ayer (1946) have argued that ethical or moral propositions are nothing more than "ejaculations" (p. 103):

> *The propositions which describe the phenomena of moral experience, and their causes, must be assigned to the science of psychology, or sociology. The exhortations to moral virtue are not propositions at all, but ejaculations or commands which are designed to provoke the reader to action of a certain sort. Accordingly, they do not belong to any branch of philosophy or science.*

Whether on religious, consistency, utilitarian, or other grounds, most common people would be taken aback by such a stance, but ideas such as these have diffused into many areas of intellectual life, oftentimes under the terms *postmodernism* or *critical theory*. No longer are books *great books*; they are merely *texts* to be analyzed and abstracted. One view is as good as another, and nonacademics are surprised to learn that intellectual life has become what appears from the outside to be merely a game of words. And once one gets the hang of the "theory" game, there is a simplicity and internal consistency about it that is no doubt the source of its attraction. But, when one examines this line of philosophical attack on its own terms—analyzing its "terms" and "texts," such reasoning may be viewed largely as tool for overturning widely shared views on almost any topic. In the limiting case, nothing is true and nothing can be known, except on an individual by individual basis.

As practical people, engineers will find such a state of intellectual affairs—so-called *solipsism*—largely revolting. Engineering is a social enterprise that depends on both an individual and a shared sense of right and wrong. Engineers commit themselves to building products, services, and organizations that *work*. They commit themselves to codes of professional conduct whether they work for an employer or directly for a paying client. To take ethics or right conduct off the table is not an option for the practicing engineer or the larger public served by his or her deeds. Academics may find such word play attractive, but if engineers themselves find it acceptable to dismiss engineering ethics as "mere ejaculations" people will die in the ensuing cesspool of shoddy work, lack of accountability, and failure. This unacceptable state of affairs takes us back to a position where we must find a way for a large group of individuals to share notions of right and wrong if we are to act together in some reasonably organized manner. This begs us to synthesize the key notions discussed above into a more integrated whole.

8.2.3   An Engineer's Synthesis of Ethical Theory

Moral theory, like much of philosophy, is a competition of ideas pursued to their logical ends. The way to make a name in philosophy is to concoct a plausible theory and pursue its logical consequences wherever they lead. It is much rarer to see philosophers *integrate* disparate theories into a consistent whole.

Engineers, by contrast, are constantly faced with physical theories that apply at different length and time scales, and they must make sense of them and know when to apply which theory to what situation. Perhaps the situation in philosophy should be viewed similarly. That is, perhaps different forms of philosophical argument should be applied in different circumstances. In particular, perhaps ethical theory should be viewed in dynamic systems terms, and different modes of philosophical thought should be considered in terms of *when* they came to be applied and how that mode of thought interacted with other beliefs about ethical matters.

Consider the following major categories of ethical thought:

1. Moral sense
2. Shared belief
3. Utilitarianism
4. Kantian consistency
5. Moral skepticism

A schematic of the plausible dynamic interactions of these five modes is presented in Figure 8.1. Starting from the naturalistic tradition, we assume the existence of a moral sense in humans, and from that seed, shared beliefs result in reductions in violence through loyalty among related groups of humans. This permits effective small-scale social organization; the benefits of such cooperation on a small scale lead to the recognition that social rules of conduct convey benefits to all. The utility of such rules drives their wider spread, and quite

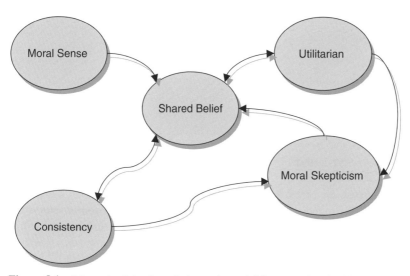

Figure 8.1   Schematic of the dynamic interactions of different modes of ethical thought.

naturally concern arises to whether those rules are consistent and universal. Over time, we grow too clever by half, elevate ethical reasoning to abstract word play, and start to question the validity of the rules that have served us so well. The resulting degradation of society leads to a reevaluation of the rising skeptical trend (largely on utilitarian and consistency grounds) and virtuous behavior rises again. It might be an interesting exercise to attempt to write equations for the evolution of the different strands of thought along evolutionary lines; however, our purpose here was merely to offer a systems-level view of the interaction of different strands of thought in one plausible manner. The more important issue for us is to get from ethical theory to ethical practice.

8.3   FROM ETHICAL THEORY TO PRACTICE

Ethical theory is one thing, ethical practice is another, and what makes the practice side of things so difficult? Simply stated, staying on the ethical straight and narrow, whether personally or professionally, requires knowing when a moral crossroads has been reached, determining the right thing to do, and doing it. Upon rare occasion, we come to a difficult ethical dilemma that requires careful thought before action, and in those few cases, the knowing-we-have-a-problem and knowing-what-to-do part of things is hard; however, the more usual case is that we are faced with a situation where there is no dilemma, not a single question, and yet we fail to do the right thing. In these *failures of responsibility* (Harris, Pritchard, & Rabins, 1995) we knew what we should have done, and yet we didn't do it. How does this happen? There are, of course, many ways to avoid doing the right thing—and most of us are well practiced in most of them—but here we consider the big three:

1. Self-interest
2. Obedience to authority
3. Conformity to the group

Each of these needs to be considered further.

8.3.1 . Self-Interest

As the comic strip character Pogo once said, "Yup son, we have met the enemy and he is us." When we know what the right thing to do is and we don't do it, one of the most likely culprits is our own self-interest. Oftentimes, moral dilemmas present us with a choice between a behavior that is attractive (and not right) and one that is difficult (and the morally correct course of action). In those cases when the attractive and incorrect behavior wins out, we often find a way of *rationalizing* our choice. We argue that the behavior we chose wasn't really wrong, didn't really hurt anyone, or that the correct course of action was too difficult or impossible. Such choices are always difficult, but learning to listen to and then ignore the little internal voice of rationalization can help us do the right thing on a more regular basis.

8.3.2 Obedience to Authority

We are often ethically defeated by ourselves, but *obedience to authority* is another important source of difficulty. In 1963, Stanley Milgram performed an important set of psychological experiments at Yale University. In these experiments, subjects were told that they would participate in an experiment on learning. The subject was asked to administer a set of shocks as a punishment for incorrect learning of a memory task, increasing in severity from 15 to 450 volts, to another "subject" who in fact was Stanley Milgram's assistant, an actor in his 50s. Although there was no real electric shock administered, the subject believed it to be real because a real, albeit small, demonstration shock was given to the subject prior to starting the "learning experiment."

During the tests, the actor playing the learner would start to complain verbally about the pain at 150 volts, complain about his bad heart at 250 volts, and kick the wall and go silent at 300 volts. During the tests, an "experimenter" in a white lab coat would stand by and calmly encourage the subject to continue the experiment. In all, 40 subjects were tested in this manner, and 26 of 40 went all the way to maximum shock level and all of the subjects went to at least 300 volts.

The level of obedience surprised even Milgram, and these experiments are often cited to explain atrocities in war and elsewhere. In an engineering context, the consequences of obedience to authority are usually less directly and obviously harmful, but the social setting is similar. Encouragement to do something unprofessional or unethical can often come from someone in authority, and the Milgram experiments show fairly conclusively that saying no to authority can require unusual courage.

8.3.3 Conformity to the Group

Blindly obeying authority can lead us astray, but so can conforming to a group. In 1951, Solomon Asch asked subjects to participate in a study of visual perception judging the relative lengths of lines. They were shown cards such as that in Figure 8.2 and asked to choose which of the three lines on the right is the same length as the line on the left. Of course, the task is obvious and individually subjects chose the right answer. Then Asch asked the subjects to do the same task in groups of 8 to 10, where the other "subjects" were confederates of the experimenter and were instructed to answer incorrectly (and unanimously) in 12 of 18 of the trials. Asch arranged for the real subject to be the next-to-last person to answer in the group. Asch thought that most people would resist the group and answer correctly.

To Asch's surprise, 37 of the 50 subjects conformed wrongly to the majority at least once, and 14 subjects conformed on half or more of the 12 incorrect trials. Other experiments have confirmed the conformity result, but people go along with the group for one of two reasons: They want to be liked by the group or because they believe the group is better informed.

Regardless of the reasons behind the conformity, the ramifications for an engineering context are important. Ethical dilemmas arise for engineers almost

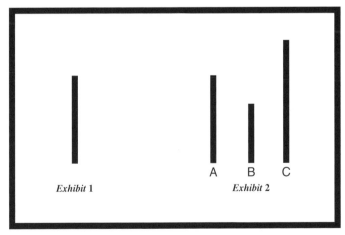

Figure 8.2 In Asch's 1951 experiments to test conformity to a group, subjects were shown a series of cards and asked to match the line on the left with the line of nearest size on the right. When performed by an individual, the task is performed correctly. When confederates intentionally give a wrong answer prior to the subject's answer, subjects often conform and give the wrong answer.

always in a group context. If there is any tendency for a group to take the ethically easy way out of a difficult situation, this will make it harder for an individual to come forward and do the right thing. Moreover, the social dynamic of the group may be compounded by having to go up against one or more authority figures at the same time. Seen in this way, it is a wonder that whistles are ever blown in hierarchical organizations of any size.

> ### *Exploration Exercise*
>
> Consider a particular business or engineering scandal in the media and analyze whether the wrongdoing was intentional or not, and whether it involved self-interest, obedience to authority, or conformity to the group. Being as specific as possible, write a short essay identifying the key parties to the scandal, their roles in perpetrating the scandal, whether or not their actions appeared conscious or unintended, and the role of self-interest, obedience, or conformity to the sequence of events.

8.3.4   Practice Makes Perfect

With self-interest, obedience to authority, and conformity to the group lurking to distort our moral compasses at many turns, doing the right thing is hard. In many engineering ethics courses, the usual approach is to work from the top down by studying the big stuff first. Read case studies of the *Challenger* or the Bhopal disasters. And indeed, we want to discuss some of these larger matters, but these larger situations are not representative of the garden variety of ethical matter that will face the usual engineering graduate. Ethics with a capital E is misleading in that the situations are enormously complex, relatively infrequent, and the individual engineer's role in the mishap is usually submerged. Exercising ethical judgment is like other complex skills—it requires plenty of practice—but if we spend our time thinking mainly about hitting the ethical homerun, we're going to swing and miss during everyday batting practice. Therefore, when we are discussing ethical issues, perhaps it is best to start a little closer to home.

If we are studying ethical issues in school, perhaps it would be better to start with discussions of cheating on homework and examinations, the purchasing of term papers, plagiarism, and more garden-variety types of ethical missteps. If we are discussing ethics on the Web, perhaps we should discuss downloading and sharing of music or software files without permission. If we are already in the business world, perhaps it would be best to discuss the padding of expense accounts, the theft of office supplies from an employer, not giving a full day's work for a full day's pay, the presentation of another's idea as your own, and the like.

If we confront these everyday matters and develop an ethical approach toward them, when bigger issues arise, we will have practice doing right when the stakes were smaller and fewer people were watching. That practice doesn't guarantee we will do the right thing when the stakes are high, but exercising our moral muscles will help us in tight spots when our self-interest, obedience to authority, and conformity to the group are conspiring to get us to go against the actions we know are right.

As with other complex disciplines, the way to improve is through practice in matters small and large. More generally, in approaching ethical questions in engineering design, in manufacturing, in business, or in life, it is better to think of flesh-and-blood human beings. Rather than thinking about anonymous customers, we should think about clients with names, voices, and faces; or we should think about our families and our neighbors—real spouses and real kids.

Exploration Exercise

For one or two days, keep an ethics-courtesy log where you record your activities along with a brief analysis of the ethical implications of those activities. Many ordinary activities have an ethical component. Do you reflexively tell the truth, or do you make up little stories? Do you use registered software, music, and other intellectual property, or do you download or use copies without cost? Are you courteous to others by saying "please" and "thank you," or do you order people around, issuing commands when you want something? Write a short essay reflecting on your logging experience and the content of this section.

8.4 FROM PERSONAL TO ENGINEERING ETHICS

An underlying theme of the present approach is that engineering ethics should be viewed as largely an extension of personal ethics. In that view, it is important to understand key points of ethical theory and practice in everyday life before adding the complexities of the engineer's world to the mix. As engineers are professionals who combine an unholy mix of business and science/technology in practice, it is not surprising that engineering ethics is itself an unholy mix of business and professional ethics together with the ethics of science and technology thrown in for good measure.

To better understand this mixture, we should start by understanding the notion of a professional, consider a number of engineering codes of ethics, examine the notion of a conflict of interest, and think about whistleblowing and its consequences.

8.4.1 What Is a Profession?

It is often stated that engineers are members of a learned profession, and a number of authors have tried to characterize the attributes of a profession. Many of these use Greenwood's 1957 five attributes of a profession as a starting point (Greenwood, 1957):

1. A systematic body of knowledge
2. Professional authority and credibility

3. Regulation and control of members

4. A professional code of ethics

5. A culture of values, norms, and symbols

In terms of these attributes, engineering appears to measure up on items 1, 2, and 5, and engineering codes of ethics require our further study; however, the regulation and control of members is a weak link, especially if we compare engineering to other professions such as medicine and law. In those other professions, the very terms "physician" and "lawyer" are controlled by licensing and professional organizations, whereas in engineering, the term "engineer" can be used to describe locomotive drivers and custodians as well as those of us who design and build technological artifacts.

The tug-of-war between professionalism and public service on the one hand and serving the interests of large-scale business interests has shaped the social history of engineering in important ways (Layton, 1990), and it is likely to continue doing so for the foreseeable future. As one examines different codes of ethics, we will see different emphases depending on whether a particular professional society is weighted toward a business or a professional model of engineering practice.

8.4.2   A Tale of Two Codes

Engineering is a divided profession, and different codes of ethics govern different engineers. Here we compare two very different codes of ethics:

1. National Society for Professional Engineering (NSPE)

2. Institute for Electrical and Electronics Engineers (IEEE)

A comparison of the NSPE (2003) and IEEE codes is a study of extremes. Where the NSPE code is lengthy, precise, and detailed, the IEEE is terse, vague, and general. Of course, the NSPE code is designed to govern the practice of engineers who work as professionals, and the IEEE code is aimed at an audience that is largely unregistered and employed by corporations.

NSPE Code of Ethics for Engineers
Reprinted by permission of the National Society of
Professional Engineers (NSPE) www.nspe.org.

Preamble

Engineering is an important and learned profession. As members of this profession, engineers are expected to exhibit the highest standards of honesty and integrity. Engineering has a direct and vital impact on the quality of life for all people. Accordingly, the services provided by engineers require honesty, impartiality, fairness, and equity, and must be dedicated to the protection of the public health, safety, and welfare. Engineers must perform under a standard of professional behavior that requires adherence to the highest principles of ethical conduct.

I. Fundamental Canons

Engineers, in the fulfillment of their professional duties, shall:

1. Hold paramount the safety, health and welfare of the public.
2. Perform services only in areas of their competence.
3. Issue public statements only in an objective and truthful manner.
4. Act for each employer or client as faithful agents or trustees.
5. Avoid deceptive acts.
6. Conduct themselves honorably, responsibly, ethically, and lawfully so as to enhance the honor, reputation, and usefulness of the profession.

II. Rules of Practice

1. Engineers shall hold paramount the safety, health, and welfare of the public.
   a. If engineers' judgment is overruled under circumstances that endanger life or property, they shall notify their employer or client and such other authority as may be appropriate.
   b. Engineers shall approve only those engineering documents that are in conformity with applicable standards.
   c. Engineers shall not reveal facts, data, or information without the prior consent of the client or employer except as authorized or required by law or this Code.
   d. Engineers shall not permit the use of their name or associate in business ventures with any person or firm that they believe are engaged in fraudulent or dishonest enterprise.
   e. Engineers shall not aid or abet the unlawful practice of engineering by a person or firm.
   f. Engineers having knowledge of any alleged violation of this Code shall report thereon to appropriate professional bodies and, when relevant, also to public authorities, and cooperate with the proper authorities in furnishing such information or assistance as may be required.
2. Engineers shall perform services only in the areas of their competence.
   a. Engineers shall undertake assignments only when qualified by education or experience in the specific technical fields involved.
   b. Engineers shall not affix their signatures to any plans or documents dealing with subject matter in which they lack competence, nor to any plan or document not prepared under their direction and control.
   c. Engineers may accept assignments and assume responsibility for coordination of an entire project and sign and seal the engineering documents for the entire project, provided that each technical segment is signed and sealed only by the qualified engineers who prepared the segment.
3. Engineers shall issue public statements only in an objective and truthful manner.
   a. Engineers shall be objective and truthful in professional reports, statements,

or testimony. They shall include all relevant and pertinent information in such reports, statements, or testimony, which should bear the date indicating when it was current.

b. Engineers may express publicly technical opinions that are founded upon knowledge of the facts and competence in the subject matter.

c. Engineers shall issue no statements, criticisms, or arguments on technical matters that are inspired or paid for by interested parties, unless they have prefaced their comments by explicitly identifying the interested parties on whose behalf they are speaking, and by revealing the existence of any interest the engineers may have in the matters.

4. Engineers shall act for each employer or client as faithful agents or trustees.

a. Engineers shall disclose all known or potential conflicts of interest that could influence or appear to influence their judgment or the quality of their services.

b. Engineers shall not accept compensation, financial or otherwise, from more than one party for services on the same project, or for services pertaining to the same project, unless the circumstances are fully disclosed and agreed to by all interested parties.

c. Engineers shall not solicit or accept financial or other valuable consideration, directly or indirectly, from outside agents in connection with the work for which they are responsible.

d. Engineers in public service as members, advisors, or employees of a governmental or quasi-governmental body or department shall not participate in decisions with respect to services solicited or provided by them or their organizations in private or public engineering practice.

e. Engineers shall not solicit or accept a contract from a governmental body on which a principal or officer of their organization serves as a member.

5. Engineers shall avoid deceptive acts.

a. Engineers shall not falsify their qualifications or permit misrepresentation of their or their associates' qualifications. They shall not misrepresent or exaggerate their responsibility in or for the subject matter of prior assignments. Brochures or other presentations incident to the solicitation of employment shall not misrepresent pertinent facts concerning employers, employees, associates, joint venturers, or past accomplishments.

b. Engineers shall not offer, give, solicit or receive, either directly or indirectly, any contribution to influence the award of a contract by public authority, or which may be reasonably construed by the public as having the effect of intent to influencing the awarding of a contract. They shall not offer any gift or other valuable consideration in order to secure work. They shall not pay a commission, percentage, or brokerage fee in order to secure work, except to a bona fide employee or bona fide established commercial or marketing agencies retained by them.

III. Professional Obligations

1. Engineers shall be guided in all their relations by the highest standards of honesty and integrity.

a. Engineers shall acknowledge their errors and shall not distort or alter the facts.

b. Engineers shall advise their clients or employers when they believe a project will not be successful.

c. Engineers shall not accept outside employment to the detriment of their regular work or interest. Before accepting any outside engineering employment they will notify their employers.

d. Engineers shall not attempt to attract an engineer from another employer by false or misleading pretenses.

e. Engineers shall not promote their own interest at the expense of the dignity and integrity of the profession.

2. Engineers shall at all times strive to serve the public interest.

   a. Engineers shall seek opportunities to participate in civic affairs; career guidance for youths; and work for the advancement of the safety, health, and well-being of their community.

   b. Engineers shall not complete, sign, or seal plans and/or specifications that are not in conformity with applicable engineering standards. If the client or employer insists on such unprofessional conduct, they shall notify the proper authorities and withdraw from further service on the project.

   c. Engineers shall endeavor to extend public knowledge and appreciation of engineering and its achievements.

3. Engineers shall avoid all conduct or practice that deceives the public.

   a. Engineers shall avoid the use of statements containing a material misrepresentation of fact or omitting a material fact.

   b. Consistent with the foregoing, engineers may advertise for recruitment of personnel.

   c. Consistent with the foregoing, engineers may prepare articles for the lay or technical press, but such articles shall not imply credit to the author for work performed by others.

4. Engineers shall not disclose, without consent, confidential information concerning the business affairs or technical processes of any present or former client or employer, or public body on which they serve.

   a. Engineers shall not, without the consent of all interested parties, promote or arrange for new employment or practice in connection with a specific project for which the engineer has gained particular and specialized knowledge.

   b. Engineers shall not, without the consent of all interested parties, participate in or represent an adversary interest in connection with a specific project or proceeding in which the engineer has gained particular specialized knowledge on behalf of a former client or employer.

5. Engineers shall not be influenced in their professional duties by conflicting interests.

   a. Engineers shall not accept financial or other considerations, including free engineering designs, from material or equipment suppliers for specifying their products.

    b. Engineers shall not accept commissions or allowances, directly or indirectly, from contractors or other parties dealing with clients or employers of the engineer in connection with work for which the engineer is responsible.

6. Engineers shall not attempt to obtain employment or advancement or professional engagements by untruthfully criticizing other engineers, or by other improper or questionable methods.

    a. Engineers shall not request, propose, or accept a commission on a contingent basis under circumstances in which their judgment may be compromised.

    b. Engineers in salaried positions shall accept part-time engineering work only to the extent consistent with policies of the employer and in accordance with ethical considerations.

    c. Engineers shall not, without consent, use equipment, supplies, laboratory, or office facilities of an employer to carry on outside private practice.

7. Engineers shall not attempt to injure, maliciously or falsely, directly or indirectly, the professional reputation, prospects, practice, or employment of other engineers. Engineers who believe others are guilty of unethical or illegal practice shall present such information to the proper authority for action.

    a. Engineers in private practice shall not review the work of another engineer for the same client, except with the knowledge of such engineer, or unless the connection of such engineer with the work has been terminated.

    b. Engineers in governmental, industrial, or educational employ are entitled to review and evaluate the work of other engineers when so required by their employment duties.

    c. Engineers in sales or industrial employ are entitled to make engineering comparisons of represented products with products of other suppliers.

8. Engineers shall accept personal responsibility for their professional activities, provided, however, that engineers may seek indemnification for services arising out of their practice for other than gross negligence, where the engineer's interests cannot otherwise be protected.

    a. Engineers shall conform with state registration laws in the practice of engineering.

    b. Engineers shall not use association with a nonengineer, a corporation, or partnership as a "cloak" for unethical acts.

9. Engineers shall give credit for engineering work to those to whom credit is due, and will recognize the proprietary interests of others.

    a. Engineers shall, whenever possible, name the person or persons who may be individually responsible for designs, inventions, writings, or other accomplishments.

    b. Engineers using designs supplied by a client recognize that the designs remain the property of the client and may not be duplicated by the engineer for others without express permission.

    c. Engineers, before undertaking work for others in connection with which the engineer may make improvements, plans, designs, inventions, or other records that may justify copyrights or patents, should enter into a positive agreement regarding ownership.

    d. Engineers' designs, data, records, and notes referring exclusively to an employer's work are the employer's property. The employer should indemnify the engineer for use of the information for any purpose other than the original purpose.

    e. Engineers shall continue their professional development throughout their careers and should keep current in their specialty fields by engaging in professional practice, participating in continuing education courses, reading in the technical literature, and attending professional meetings and seminars.

— As Revised January 2006

"Engineers shall strive to adhere to the principles of sustainable development[1] in order to protect the environment for future generation."

Statement by NSPE Executive Committee

In order to correct misunderstandings which have been indicated in some instances since the issuance of the Supreme Court decision and the entry of the Final Judgment, it is noted that in its decision of April 25, 1978, the Supreme Court of the United States declared: "The Sherman Act does not require competitive bidding."

    It is further noted that as made clear in the Supreme Court decision:

1. Engineers and firms may individually refuse to bid for engineering services.

2. Clients are not required to seek bids for engineering services.

3. Federal, state, and local laws governing procedures to procure engineering services are not affected, and remain in full force and effect.

4. State societies and local chapters are free to actively and aggressively seek legislation for professional selection and negotiation procedures by public agencies.

5. State registration board rules of professional conduct, including rules prohibiting competitive bidding for engineering services, are not affected and remain in full force and effect. State registration boards with authority to adopt rules of professional conduct may adopt rules governing procedures to obtain engineering services.

6. As noted by the Supreme Court, "nothing in the judgment prevents NSPE and its members from attempting to influence governmental action. . ."

NOTE: In regard to the question of application of the Code to corporations vis-à-vis real persons, business form or type should not negate nor influence conformance of individuals to the Code. The Code deals with professional services, which services must be performed by real persons. Real persons in turn establish and implement policies within business structures. The Code is clearly written to apply to the Engineer and items incumbent on members of NSPE to endeavor to live up to its provisions. This applies to all pertinent sections of the Code.

---

[1] *Sustainable development* is the challenge of meeting human needs for natural resources, industrial products, energy, food, transportation, shelter, and effective waste management while conserving and protecting environmental quality and the natural resource base essential for future development.

The NSPE code is detailed to the point where it offers practicing professionals practical guidance in creating an ethical engineering practice. The NSPE as an organization takes ethical matters seriously and regularly reviews ethics cases with its Board of Ethical Review (BER). Studying NSPE cases and their review can be a useful way to see how the code applies in practice. Cases are available at www.nspe.org or at a number of other online engineering ethics sites.

The IEEE code (1990) is a study in contrasts. It is as brief as the NSPE code is long. It is general where the NSPE code is specific. Of course, most of the IEEE's members are not in private practice, but instead work for large corporations.

IEEE Code of Ethics
(Reprinted by permission of the IEEE)

We, the members of the IEEE, in recognition of the importance of our technologies in affecting the quality of life throughout the world, and in accepting a personal obligation to our profession, its members and the communities we serve, do hereby commit ourselves to the highest ethical and professional conduct and agree:

1. to accept responsibility in making engineering decisions consistent with the safety, health and welfare of the public, and to disclose promptly factors that might endanger the public or the environment;

2. to avoid real or perceived conflicts of interest whenever possible, and to disclose them to affected parties when they do exist;

3. to be honest and realistic in stating claims or estimates based on available data;

4. to reject bribery in all its forms;

5. to improve the understanding of technology, its appropriate application, and potential consequences;

6. to maintain and improve our technical competence and to undertake technological tasks for others only if qualified by training or experience, or after full disclosure of pertinent limitations;

7. to seek, accept, and offer honest criticism of technical work, to acknowledge and correct errors, and to credit properly the contributions of others;

8. to treat fairly all persons regardless of such factors as race, religion, gender, disability, age, or national origin;

9. to avoid injuring others, their property, reputation, or employment by false or malicious action;

10. to assist colleagues and co-workers in their professional development and to support them in following this code of ethics.

Approved by the IEEE Board of Directors
August 1990

Exploration Exercise

Compare and contrast the IEEE and NSPE codes of ethics in a short essay. In what ways are the two codes similar and in what ways are they different. Consider the nature of the two organizations and hypothesize as to the nature of the differences.

8.4.3   Conflicts of Interest

Almost all codes of ethics warn against conflict of interest, but there is a substantial amount of confusion over what constitutes a conflict of interest, and many codes, interestingly enough, do not define them directly. For example item 2 of the IEEE code has the following to say:

> 2. *avoid real or perceived conflicts of interest whenever possible, and to disclose them to affected parties when they do exist;*

Apparently conflicts are something to be avoided, but the IEEE code is not helpful in defining them. The NSPE code addresses conflicts of interest under the rubric of "faithful agency" it gives several examples in item 4:

> 4. *Engineers shall act for each employer or client as faithful agents or trustees.*
>
> > **a.** *Engineers shall disclose all known or potential conflicts of interest that could influence or appear to influence their judgment or the quality of their services.*
> >
> > **b.** *Engineers shall not accept compensation, financial or otherwise, from more than one party for services on the same project, or for services pertaining to the same project, unless the circumstances are fully disclosed and agreed to by all interested parties.*
> >
> > **c.** *Engineers shall not solicit or accept financial or other valuable consideration, directly or indirectly, from outside agents in connection with the work for which they are responsible.*
> >
> > **d.** *Engineers in public service as members, advisors, or employees of a governmental or quasi-governmental body or department shall not participate in decisions with respect to services solicited or provided by them or their organizations in private or public engineering practice.*
> >
> > **e.** *Engineers shall not solicit or accept a contract from a governmental body on which a principal or officer of their organization serves as a member.*

Item a mentions the term "conflicts of interest" and b to e are examples of classical conflicts of interest, but the NSPE code is little help in defining what we mean by a conflict of interest.

In fact, there is a good deal of confusion over conflicts of interest, and we must be much clearer about what one isn't and is. The first point to make is that a conflict of interest is not any ordinary ethical lapse. If you are dishonest, betray a confidence, intentionally do substandard work, work outside your area of confidence, or otherwise behave in a typically unprofessional manner, you may have acted unethically, but you have not done so in the face of or as a result of a conflict of interest.

No, conflicts of interest do not (cannot) arise in simple two-sided relationships where one party plays a single role with respect to another. Conflicts of interest arise when a party plays multiple roles between two or more parties, and may be defined concisely as follows:

> A **conflict of interest** *arises when a party may be forced to choose between the obligations of two or more roles in a manner where the interests of one party may be elevated over the interests of another.*

Each of these terms deserves brief comment followed by a diagram and some discussion of the canonical conflict of interest that faces all practicing engineers.

The term *party* here is meant to represent an individual, organization, or aggregate of individuals, and this might be an engineer, a company, a client, or some other party to an ethical transaction. The term *role* indicates what relationship a party has with another; for example, an engineer may be an employee or a consultant to a company. Roles carry certain "obligations" of satisfactory performance. Engineering employees and consultants, for example, are expected to create effective designs for those who employ them, and in this way we see that the obligations of various roles are related to the interests of the related party.

Of course, all this sound fairly confusing, but it really is quite simple, as we can see in Figure 8.3. In this figure, what we call a *party–role* (PR) *diagram,* we see the canonical parties and roles in a typical engineer's life. The PR diagram is drawn with respect to the conflicted party, and the roles are shown as directed arrows away from the conflicted (or potentially conflicted) party. In the particular drawing, we see how engineers typically work as *consultants* for *clients* or

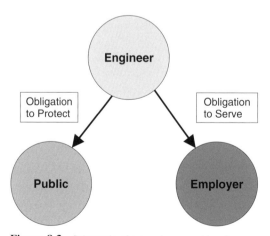

Figure 8.3   Schematic of an engineer working for an employer who also has an obligation to the public. Such party–role diagrams are helpful in understanding and visualizing conflicts of interest.

employees for *companies*. In addition, engineers—by virtue of their professional status—have an obligation to serve the general good of *society* or the *public* as a *guardian* or *protector*. The very nature of these relationships opens the door that the obligations of the two roles may sometimes collide.

For example, if your boss gives you an order to pour a carcinogenic solvent down a sanitary sewage drain, you are being asked to make a choice between the obligation you have in your role as employee and keep your obligation to follow the directions of your boss and the obligation you have to protect the public from illegal dumping of restricted substances. Of course, in this example, the right thing to do is clear, but understanding conflict of interest more generally, the different roles you play, and the different obligations you face may help you understand difficult ethical decisions more thoroughly.

Exploration Exercise

You work for ABC Consultants as a project engineer and you have a choice of buying a ticket to Atlanta, Georgia, using company funds for $500 on AAA Airlines or $250 on FlyCheap Airways. You have a frequent flyer plan with AAA but not FlyCheap. In a short paragraph, consider the conflict of interest you have in this setting. Describe what you believe to be the ethical course of action and why. Draw a PR diagram sketching the basic conflict of interest, labeling the parties and roles clearly. Detail the obligations of the different roles in words besides the sketch. Suppose that ABC Consultants requires all frequent flier miles collected in the course of company business to be turned over to the company. Draw the resulting PR diagram and explain how this change in rules affects the conflict of interest.

8.4.4 Whistleblowing Is Not a First Resort

Whistleblowing gets a lot of attention in the press, but going public with objections to your employer's policies, plans, or products may get you fired. Whistleblowing laws exist in some states, but much employment in the United States remains governed by the doctrine of "at-will employment," meaning that an employee can be terminated at any time for any reason. Your best protection against ever having to face a choice between going along with an unethical policy or decision or losing your job is to seek employment with organizations (a) known for products and services in line with your values and (b) who have reputations for ethical conduct. If you thought you joined such a company, but repeatedly find yourself in ethically uncomfortable situations, this may be an important signal to look for a new job.

Having said this, what can you do if, as part of your work, you are confronted with something you believe is very wrong? The first thing to do is to make sure that you believe the matter of concern *is* very serious. If you do, you must then bring it to the attention of others above you in the hierarchy. Most often this

would require you to approach your immediate supervisor, but there may be circumstances where that may not be wise. Perhaps there are co-workers who agree
and who will stand by you. Perhaps there are other managers who understand
the problem and are willing to argue your position. If your organization has an
ethics hotline or ombudsman charged with looking into ethical concerns, it may
be appropriate to use those channels.

If after going through channels inside the organization the matter is not
resolved, you may be forced to go to the press or to an appropriate governmental
agency. This is not a step to be taken lightly, and even with the best of whistle-
blower laws, whistleblowers pay a heavy personal price to set the ethical record
straight.

SUMMARY

We started the chapter by examining how engineering ethics is often perceived by engineering students as a fairly boring and unimportant topic. The importance of the subject is
underlined by widespread ethical lapses in business and public life, and the profession of
engineering, too, is subject to these difficulties. As we discovered, the subject needn't be
boring if (a) we approach it with an engineer's eye for modeling and (b) work from the
bottom up to hone our professional ethical reflexes on top of the personal ethical notions
we learned at home, in church, at school, or in our communities at large.

Most of us gain entryway into the realm of moral reasoning from one or more golden
rules, and here we examined a number of golden rules from a number of the world's
great religions and cultures. We found that the rules come in two flavors, positive and
negative golden rules. Negative golden rules ask us to behave consistently, not doing unto
others what we would not have done to ourselves. Positive golden rules ask us to behave
charitably, holding us up to an ideal of good deeds and works.

Although golden rules are a useful entry point for discussion of morality, they are not
the end of the story because golden rules are silent on what constitutes wrong or right
conduct. This has led philosophers over the centuries to contemplate different theories
of ethics, and here we very briefly considered five of them, but one of the key points
was to understand their relationship, one to the other, and how ethical understanding is
something of a patchquilt of different modes of reasoning, in much the same way that
engineering model building often uses different models in different circumstances.

Theorizing is one thing, but engineers' interest in theory is putting it to practice. In that
vein, we recognized that oftentimes engineers know the right thing to do, but sometimes
they choose not to do it. Three obstacles to responsible behavior, self-interest, obedience
to authority, and conformity to the group, have been discussed, and relevant results from
social psychology experiments have been discussed. Understanding these results helps
us recognize that doing the right thing, especially in an organizational setting, can be
a terrifically difficult matter. A key response to this realization has been to advocate
practicing ethical behavior in the small things everyday. In this way, when a larger ethical
issue arises, the engineer will be better prepared to do the right thing when the stakes are
higher, many are watching, and the pressure to do wrong is at its worst.

Finally, this led us to consider how engineering ethics necessarily goes beyond personal
ethics, because of the complexity of engineering social interaction, the dictates of professional practice, and the existence of explicit engineering codes of ethics. After reviewing
two engineering codes of ethics, we considered how conflicts of interest arise between

the obligations of different roles between two or more parties. Thereafter, we considered how engineers sometimes may be called on to stand up within their organization and do the right thing, despite resistance from their co-workers and supervisors. In the extreme case, whistleblowing, going public to the press or government with details of wrongdoing may be necessary, but whistleblowing is an extreme step, and it should only be used as a last resort, after other methods of influencing the organization have been exhausted.

EXERCISES

1. Read about a major corporate scandal or engineering disaster. In a short paragraph, identify key parties to the mishap and identify ways in which ethical wrongdoing led to trouble.

2. Make a list of 10 things that depend in part for their existence on social agreement.

3. Write a short essay on the sources of or influences on your own sense of right or wrong. Did this come from your parents, peers, religious training, or where? How do you personally determine right from wrong?

4. Current surveys suggest that large numbers of college students cheat on examinations. Is cheating ethically acceptable or unacceptable? Write a short paragraph explaining why or why not.

5. Current surveys suggest that many students download music they did not purchase from the Web. Is such downloading ethically acceptable? Write a short paragraph explaining why or why not?

6. Imagine you are a talented rock musician with a number of popular CDs. Reconsider your response to Exercise 5. Did your response change depending upon your point of view?

7. Lawrence Kohlberg articulated a theory of moral development with three levels of development and two stages within each level. Using the web, investigate the levels and stages of Kohlberg's ladder, and write a short paragraph whether there is a connection between any of the stages and the major modes of moral thinking discussed herein.

8. Look up the Association for Computing Machinery (ACM) code of ethics at www.acm.org. Compare and contrast the ACM code with that of the IEEE in a short essay.

9. Consider a company in the news because of ethical missteps. Investigate whether that company has a code of ethics, and determine which elements of the company's code were violated in the mishap.

10. Consider a company you admire that also has a code of ethics. In a short essay discuss the code of ethics and the ways in which it reflects or does not reflect what you know about the company.

11. Scientific organizations sometimes have codes of ethics. Look up the code of ethics of a scientific organization and compare and contrast that code to the NSPE code of ethics in the text.

12. Some items in engineering codes of ethics have less to do with ethical matters and more to do with regulating engineering commerce and trade. Examine the NSPE code of ethics and determine which items are of this nature. In a short paragraph discuss whether you view these as ethical issues or not.

13. NSPE case studies are available on the Web (www.nspe.org). Read a case study involving conflict of interest and draw a clear PR diagram labeling parties and roles. In words, write the obligations of the roles that were at issue in the conflict.

14. M. W. Thring's book, *The Engineer's Conscience* (London: Northgate, 1980), contains six propositions that he says are "necessary conditions for the survival of civilization." They are as follows:

- There is only one humane way of leveling off the world's population, and this is to provide a fully adequate standard of living and education to all people.

- The enormous differences in standard of living and use of resources between groups of people must be essentially eliminated within one generation if we are to eliminate the tensions leading to World War III.

- No pollutant must be emitted to the atmosphere, to water, or to land until it has been proved conclusively that the level of pollution has no long-term harmful cumulative effect on people, animals, or plants.

- It is a necessary condition for a stable civilization in the next century that the rich countries gradually eliminate their nonproductive activities, such as advertising, weapons manufacture, and fashion and built-in obsolescence, and replace these with genuine attempts to help the poor countries to build up the equipment and knowledge to become full self-supporting at a good standard of living.

- We have to bring about a fundamental change in the ethos of our society if it is to have any chance of moving into a stable 21st-century world.

Write an essay debating the merits of any one of Thring's "propositions."

Chapter 9

Pervasive Teamwork

9.1 OUR LOVE–HATE RELATIONSHIP WITH TEAMS

The modern literature of management is full of teams and so is the modern workplace. Much of the current emphasis on teams can be traced directly to the success of Japanese companies such as Toyota in using teamwork as part of various *quality methods* that borrowed heavily from ideas introduced by the American W. Edwards Deming as part of the postwar rebuilding of Japanese industry (Scholtes, 1998). The advantages of teamwork are many: Effective teams bring together the complementary skills needed to do a job; they can make better decisions than an isolated individual; and they can effect better, more efficient execution than is possible with a more loosely knit group of individuals.

Interestingly, the modern love affair with teams stands in stark contrast to the unpleasant team or group experiences that many of us have experienced on *group projects.* Significant numbers of readers scanning these words have had the misfortune to be forced to single-handedly carry some number of freeloaders on some group project. Although by themselves, these less-than-positive experiences don't condemn group and team projects, they do cause us to be realistic about teams and face teamwork benefits and difficulties in a clearheaded manner.

We start by examining some of the differences between groups and teams; these distinctions lead us to examine the case for establishing team ground rules to help align team member expectations right from the start. Thereafter, we derive a number of simple quantitative models that help us understand some of the obstacles to effective teamwork. We conclude with a discussion of an effective brainstorming protocol that can help problem-solving teams be more effective in reaching high-quality solutions.

9.2 WORKING TOGETHER IN GROUPS AND TEAMS

The impulse for human beings to work together in groups is biologically irresistible. Ours is a social species, and we have hunted, gathered, farmed, and invented our way into large-scale social organization over many thousands of years. Modern organizational theory and practice has weighed in with a number

The Entrepreneurial Engineer, by David E. Goldberg
Copyright © 2006 John Wiley & Sons, Inc.

of refinements on our natural inclinations and instincts. Here we consider the differences between a group and a team, some team basics, and a number of key team ground rules.

9.2.1   Teams versus Groups: What's the Difference?

In a modern organizational context, however, the term *team* is now used somewhat loosely to describe any *group* or *assemblage* of people, but some are more careful to distinguish groups from teams. A particularly interesting study of teams (Katzenbach & Smith, 2003) studied 50 different teams in 30 companies, trying to understand common approaches of those teams that had reached a high level of performance. Their work led them to define a *working group* as follows (p. 91):

> **Working group.** *A working group consists of members who interact primarily to share information, best practices, or perspectives and to make decisions to help each individual perform within his or her area of responsibility.*

They contrasted this loose association with a team, which they defined somewhat differently (p. 45):

> **Team.** *A team is a small number of people with complementary skills who are committed to a common purpose, performance goals, and approach for which they hold themselves mutually accountable.*

Examining the two definitions carefully, the primary distinction between the two has to do with *goals* and *accountability*. In a working group, the goals of the group are typically the goals of the larger organization: There is no special purpose or extra-organizational reason for the working group. Moreover, individuals are accountable for their own work products (or the work products of those who work for them).

On the other hand, a team has a special reason for its existence (its goal or goals), and the team members are collectively responsible for the work product that all contribute to. It is important to point out that neither working groups nor teams are inherently good or bad in and of themselves. The key point is to understand when each type of organization is more appropriate. Simple committees are often working groups where individuals come together to communicate, deliberate, and then perform their individual tasks. Special projects with difficult deadlines often need teamwork, and it is important to execute the basics well to raise the performance of the team.

9.2.2   Team Basics

The key elements of becoming a team are contained in the definition. Recall that a team is

1. a small number of people
2. with complementary skills

3. who are committed to a common purpose, performance goals, and approach

4. for which they hold themselves mutually accountable.

Decision making becomes more difficult as team size grows, and this argues for keeping team size as small as possible. Of course, every team needs a full complement of skills to get its job done, and this factor argues for increasing teams to ensure that the needed skills are available.

A key element of becoming a true team is to understand and articulate common purpose, goals, and approach. A finding of the Katzenbach and Smith (2003) study is that teams form in reaction to a performance challenge and that the challenge is more important than creating a "team-oriented" environment or the particulars of some team-training exercises. A common distinction in the literature of organizational behavior is between (1) task and (2) relationship, and much has been written about when it is important to emphasize task, relationship, or both. The study findings on teamwork are clear that good team relationships grow out of the need for high performance (high task). There is no chicken-or-egg problem when it comes to how teams are formed.

Finally, mutual accountability is an important, but difficult, element of what it means to be a team. Most incentives, evaluations, and recognition in organizations are aimed at individuals, and most such goodies come from someone else (your boss, his or her boss, a committee, etc.). Mutual accountability means that (1) the team takes responsibility for its own behavior, and (2) its evaluation mechanisms are *internal* to the team and do not rely excessively on the opinions of others. The competence and trust implicit in these matters are fairly rare, and high-performance teams are elusive yet worth the pursuit.

9.2.3 Team Ground Rules and Their Enforcement

In moving beyond group work and getting to teamwork, it is helpful to have a clear set of expectations among the team members for their obligations to the group and vice versa. Thus, it is useful to hold a number of meetings early in a team's formation to reach agreement on a set of team *ground rules.* Ground rules constitute a social contract among the members of a team as to the *expected norms* of acceptable team behavior (Scholtes, Joiner, & Streibel, 2003).

Ground rules should cover those situations that are likely to arise on a regular basis:

Governance. How will decisions be reached? By majority vote, by consensus, as recommendation to the team leader, and so forth.

Attendance. If teams are worth forming and meetings are worth having, then members should commit to attending meetings unless there are legitimate reasons for an absence. Moreover, meetings should begin and end on time, and members should commit to keeping to the schedule.

Contribution and Listening to Others. Members should commit to an open exchange of ideas governed by common courtesy. Members should contribute to meetings when they have something to say, and they should listen to the contributions of others.

Assignments and Roles. Members should commit to doing their assignments between meetings according to the team schedule. Moreover, team roles such as meeting leader and secretary can be rotated or assigned as regular team duties.

Task-Specific Rules. Rules may be needed in response to the particular task or particular group of people assembled for the team. These rules should be discussed, committed to paper, and agreed upon early in the team's formation.

Rule Enforcement. How will violations of the rules be dealt with? Sometimes simply keeping public records of attendance, timely delivery of work product, and other statistics is enough to shape up those with a tendency to go astray. More direct challenges to team coherence, such as freeloading, need to be dealt with swiftly and directly. Sometimes frank team discussions with those who are not carrying their weight are enough to rectify the problem. Other times, it may be necessary to censure or expel an errant member for the good of the team.

Although developing effective teams is not easy, teams that explicitly set clear expectations by discussing and writing out ground rules are well on the way to a more positive experience.

Exploration Exercise

Draft a set of team ground rules for a real group or team that you are now a part of or once were a part of.

9.3 UNDERSTANDING THE DIFFICULTIES OF TEAMWORK

Modern organizations use teams quite liberally, and their advantages are significant; the pursuit of improved teamwork is a laudable organizational goal. Nonetheless, all work in groups and teams must face up to a simple fact:

Working in teams is more complex than working alone.

Here, I mean "complex" in a mathematical sense, and we can make some progress by considering several mathematical models that can help us understand some of the ways in which teamwork becomes harder as team size increases. In particular, we consider a little model of teamwise deciding and doing, a little model of the probability of teamwise conflict, and a number of the ramifications of these models for team selection and sizing.

9.3.1 A Little Model of Teamwise Deciding and Doing

To keep things simple, imagine that when we work alone there are two types of work:

1. Deciding what to do
2. Doing it

Let's call the time required for the first type of work T_1 and for the second type of work T_2. Overall, the time required to complete the task alone is

$$T_{\text{alone}} = T_1 + T_2 \tag{9.1}$$

When we work on a team of size n, deciding and doing are processed differently. Imagine during the deciding phase that each of the n members takes T_1 units of time to decide what he or she would do and that this thinking is presented to the others sequentially. Thereafter, a simple vote is taken to determine which idea or combination of ideas will be executed. Subsequently, in the doing phase, the n members of the team divide the doing or task time, T_2, equally. Summing the teamwise deciding and doing time components results in an equation for the total task time for a team of size n as follows:

$$T_{\text{team}} = nT_1 + \frac{T_2}{n} \tag{9.2}$$

Note that Equation (9.2) reduces to Equation (9.1) when $n = 1$.

Taking the derivative of Equation (9.2) with respect to the team size n and setting to zero results in the following equation for optimal (fastest) team, n^*:

$$n^* = \left(\frac{T_2}{T_1}\right)^{1/2} \tag{9.3}$$

For example, consider the case where $T_1 = 0.01$ and $T_2 = 0.99$. Then $n^* = (0.99/0.01)^{1/2} \approx 10$. Figure 9.1 shows the variation of total time required as a function of team size. The time required to perform the task with the optimally sized team according to this model is one-fifth of the time required by a person working alone. In other words, the speed up ($T_{\text{alone}}/T_{\text{team}}$) of working in a team is roughly five times that of working alone.

The simple model of team decision making and doing helps us think about how teams can help further or hinder the efficient accomplishment of a particular task. We should note that the model does not include the possibility of improved solution quality as a result of more people working on forming a plan. Nor does the model account for the possibility of improved commitment to a solution that

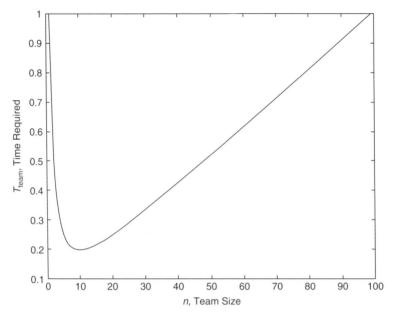

Figure 9.1   Time required for a team to complete a task depends on the decision time T_1 and the task time T_2. In the figure, task time is approximately 100 times the decision time and the optimal team size is 10.

derives from "ownership" of an idea that comes from shared decision making in a team setting. And while it might be useful to think about and derive such models, here we look at the increased pairwise relationships that come about from larger teams as a source of increased conflict.

9.3.2   A Little Model of Teamwise Conflict (and Creativity)

Increasing decision costs can overshadow the manpower advantage of teams, but another way to look at increased team size is as a source of more *relationships*. We act as individuals, but the fundamental unit of *interaction* is the *pair*. Simple combinatorial counting exercises are useful here. Since a team has n individuals, each of those can pair off with $n - 1$ of the remaining individuals, but we must be careful not to double count. Therefore, there are $n(n - 1)/2$ unique pairs among n team members. Call this number, $r(n)$, the number of relationships.

Communication among pairs of individuals can be a source of creativity and innovation, or it can be a source of conflict. Imagine we are considering the possibility of conflict, and call the quantity p the probability of a conflict between a pair of individuals. The probability that there will be conflict between in at least one relationship on a team of size n may be given by the following expression:

$$P(\text{conflict}) = 1 - (1 - p)^{r(n)} \tag{9.4}$$

Figure 9.2 shows the probability of at least one conflict on a team as a function of the pairwise probability of conflict and team size. The probability of conflict on a team will be greater than $1 - \alpha$ when $\ln 1 - \alpha > r \ln(1 - p)$. Define the critical team size, n_c, as the value that makes the inequality an equality as follows:

$$n_c = \frac{1 + \sqrt{1 + 4c}}{2} \tag{9.5}$$

where

$$c = 2 \frac{\ln 1 - \alpha}{\ln 1 - p}$$

For large n, Equation (9.5) may be approximated as $n_c = \sqrt{c}$. For small p values, Equation (9.5) may be approximated as follows:

$$n_c = \sqrt{\frac{-2 \ln 1 - \alpha}{p}} \tag{9.6}$$

The little model uses fairly simple assumptions to predict the probability of conflict. Even with fairly small relationshipwise probabilities of conflict, the probability of some team conflict grows rapidly with increased team size and suggests one of the reasons why teamwork is so difficult.

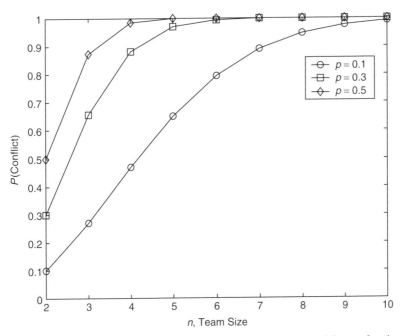

Figure 9.2 Probability of conflict (or other team property) goes up rapidly as a function of team size at fixed probability of pairwise conflict.

Interestingly, however, the same equation applies to teamwise creativity, if we simply interpret the probability p as the probability that a given pair will spark a creative idea and the probability $1 - \alpha$ as the threshold probability of teamwise creativity. Even with modest p values, relatively small teams will have high probabilities of having one creative pair, if not more.

9.4   WHY COOPERATION ISN'T EASY

One of the obstacles to effective team formation is the sheer complexity of the enterprise, but ever since kindergarten, we have been exhorted to "be cooperative" and "share" with our fellow human beings. Why isn't it easier for us to just get along, cooperate, and get the job done?

Political scientists have puzzled over this question for years, and Axelrod's idealized computer studies of the *evolution of cooperation* (Axelrod, 1984) have offered some interesting answers to this question. This is not the place to explore those studies in detail, but it is useful to look at the idealized problem of conflict and cooperation used by political scientists to dissect questions of this nature, the so-called *prisoner's dilemma problem*. The prisoner's dilemma problem is given its name because it idealizes the situation faced by a pair of criminals who have been caught by the police. In such a situation, each prisoner faces the choice of remaining silent (cooperating with his or her fellow prisoner) or taking a reduced sentence in exchange for ratting out his or her partner in crime (defecting). Many social and organizational problems may be viewed in similar terms (Miller, 1992).

A common, analogous situation in a team setting is that team members may cooperate with fellow team members and do their jobs or they may defect from the team by freeloading and letting others shoulder their burden. It is exactly this situation that causes many of us to remember group projects in less than positive terms.

Table 9.1 depicts a typical payoff matrix for the teammate's (prisoner's) dilemma. The payoff to team member 1 is listed first and that to team member 2 is listed second. Note that when both team members work, the individual rewards are high and the sum of individual awards are highest. When one teammate loafs, that individual is rewarded with more free time, credit for the finished product, but the working teammate pays the price by having to shoulder the entire burden. When faced with a situation such as this, many teammates would be tempted to loaf themselves. In the loaf–loaf case, the teammate who was working receives an increase in payoff (no longer working so hard), but the sum of the individual payoffs is minimal.

When viewed in this manner, perhaps it is less surprising that cooperation is so difficult to achieve. Many circumstances on teams and in organizations have this sort of incentive structure to them, and special efforts must be made to reach agreement among team members as to what constitutes good team and team member performance. An important element to creating high-performance teams is the holding of effective meetings.

Table 9.1 Typical Payoff Matrix for Two-Member
Teammate's Dilemma Problem

		Member 2	
		Cooperate (Work)	Defect (Loaf)
Member 1	Cooperate (Work)	(6, 6)	(0, 10)
	Defect (Loaf)	(10, 0)	(2, 2)

9.5 MEETINGS, MEETINGS, AND MORE MEETINGS

Meetings are a necessary evil in team and group life, but they need not be as evil as they often are. Meetings come in almost an infinite variety of shapes and sizes. There are regular staff meetings, one-on-one meetings, huddles, sales meetings, client meetings, crisis meetings, problem-solving meetings, brainstorming meetings, to name a few. It is important to review the three essential items that help make every meeting a success and pay special attention to problem-solving or brainstorming meetings.

9.5.1 Three Little Keys to Meeting Happiness

This may sound simple, but there are three things about meetings that are really important:

1. They should start on time.
2. They should end on time.
3. They should have an agenda.

Meetings usually involve busy people who are highly paid. Wasting just 10 minutes in starting or ending late in a weekly meeting involving 10 people wastes approximately 87 hours or over 2 person-weeks in a year. Valuing team member time at $100/hour, this totals roughly $9000 down the drain.

It is surprising the number of meetings that take place without an agenda. This is almost always a mistake. Team members want to know why they are meeting and what topics will be covered. Elaborate agendas printed in color are not necessary; a simple handwritten list of topics copied at the last minute will do in a pinch, but some planning should go into the topic selection and sequence. Of course, surprises can arise during the course of a meeting, and the meeting leader can modify the agenda on the fly if it seems prudent, but having an agenda and following a schedule are often enough to have above-average meetings and team productivity.

9.5.2 A Day in the Life of a Typical Problem-Solving Meeting

One of the most important meetings—and one of the most nettlesome—is the problem-solving or brainstorming meeting. In such meetings, teams are under the gun to come up with a solution to a nasty problem or to propose alternatives to exploit a juicy opportunity. Either way, the urgency of the situation and the need to innovate on the spot, places a premium on effective meeting structure and management. These matters have long been studied by organizational specialists, and effective meeting procedures and protocols exist, but it is surprising the number of teams and organizations that have not adopted any of these methods. Here, we probe the anatomy of a typical problem-solving session. Be warned: It is not a pretty sight, but you will recognize the problems discussed. This discussion leads naturally to a more structured approach to problem solving in a group setting. The adoption of this single technique can boost substantially the creativity, quality, and quantity of the solutions generated by even the stodgiest of organizations.

To set the stage, imagine that you've just learned of a difficult problem facing your team. Your group leader calls a meeting to discuss the difficulty, and at the appointed hour the troops gather. The high-priced talent sitting around the table knows the importance of this moment. The meeting begins. The group leader briefly outlines the difficulty and throws the meeting open for ideas. One group member begins, first raising several aspects of the difficulty that the group leader omitted from his problem description, then proposing a specific solution. A second group member challenges the effectiveness of the first member's solution on the basis of cost and also raises several other unreported aspects of the problem. She finishes her statement with her own proposed solution. A third group member finds the first two solutions inadequate on the basis of implementation time and suggests a third solution. And on it goes.

The meeting continues in this manner, with group members proposing and rejecting each other's solutions, until finally the group leader, realizing that the meeting is going nowhere, takes over. As a good leader, the best he or she can do at this point is to piece together a solution from the best suggestions made so far. A more autocratic leader simply announces the solution, a solution that the careful listener heard in the meeting opening and the leader's subsequent remarks. Either way, the group members leave with something of an empty feeling, knowing that, yet again, the group's creativity and innovative potential have gone largely untapped.

9.5.3 What's Wrong?

The scenario described above is all too familiar in business, but the reasons for the failure of most unstructured brainstorming or problem-solving meetings are fairly easy to pinpoint:

1. Inadequate discussion of the various facets of the problem
2. Premature criticism of partial solutions
3. Mixed discussion of solutions and criteria or objectives
4. Inhibition of full exploration of ideas by a leader-follower dynamic

The remainder of this section examines how each of these factors can inhibit group brainstorming productivity.

The first difficulty in unstructured problem solving is that the problem does not usually receive an adequate initial airing. In our imaginary scenario, the meeting leader started off with his view of the problem, but other people around the table obviously had additional information and different viewpoints that could have been helpful in understanding the problem more fully. Since unstructured meetings allow any kind of discussion at any time, as soon as the meeting is opened up to the group, natural human impatience almost guarantees that solutions will be proposed before the problem has received adequate discussion. This is detrimental to a group's ability to come to consensus because there is no shared vision of the range and complexity of the problem.

The usual unstructured meeting has a serial idea–criticize–idea–criticize rhythm to it, which is particularly harmful to a group's problem-solving productivity. Finding a solution to a problem usually requires the recombination of a number of notions from a number of sources to arrive at something that works. If notions are disposed of before they've had a chance either to generate or inspire other solutions or refinements, or to be recombined with other partial solutions, the end result will be less satisfactory than it otherwise might have been (i.e., had more partial solutions survived to be considered in the final analysis).

The unstructured nature of the typical meeting also leads to a mixed discussion of partial solutions along with criteria for judging those solutions. Premature rejection of partial solutions is often accomplished by raising a particular criterion by which the proposed solution is judged more or less inadequate. Of course, this disregards the other eight criteria that the solution may fully satisfy and points out the difficulty of viewing solutions in isolation with particular criteria. There are almost always trade-offs to make in choosing solutions to tough problems, and it is better to postpone the consideration of all solutions in the context of all criteria than to use individual criteria as bullets to shoot down each new idea that dares to raise its innovative little head.

The last-but-not-least difficulty in many meetings is the leader–follower dynamic. In many organizations, leaders are accustomed to making decisions without much input from those they lead; typically, a manager asks for input only to unveil the "correct" answer at the end of the "brainstorming" session. Thus many seasoned veterans come to such sessions with the idea of listening for hints about the Politburo's chosen solution rather than listening and contributing to a genuinely creative experience.

Formal managers are necessary in all organizations, and these managers have the right and duty at times to exercise their decision-making authority. When they call a brainstorming session, however, they have a moral obligation to their people to listen to their ideas and a fiduciary responsibility to corporate shareholders (or organizational backers) to try to reach the best solution possible. This requires some restructuring of the way meetings are conducted, if only to overcome the tendency of leaders to lead and followers to follow.

The four difficulties of unstructured brainstorming—inadequate discussion of the problem, premature criticism of alternatives, mixed discussion of solutions and criteria, and the workings of the leader–follower dynamic—remind us of the difficulties faced by the individual writer as he or she tries to get thoughts on paper. There, writer's block can be attributed to trying to create and criticize simultaneously. In a group problem-solving session the same conflict between creating and criticizing arises, but the size of the group complicates and intensifies the destructive reaction between the creative juices and the critical venom. Many meetings turn into survival-of-the-loudest (or longest-winded) sessions or, worse, a contest where only the biggest boss's ideas get considered. In the next section, we'll examine a structured approach to brainstorming that separates the creative and critical thinking throughout the problem-solving process, thereby permitting group productivity and creativity to flourish.

9.5.4   Structured Brainstorming

After sitting through dozens of meetings and witnessing hundreds of good ideas being shot down in the usual fashion, one begins to wonder whether there might be a better way to solve problems in a group. I know I was ripe for my first encounter with brainstorming techniques when I took a software sales training course in Indianapolis in the late 1970s. The brainstorming protocol taught was based on an adaptation of Alex Osborn's (1963) original protocol; it's been used widely at General Electric and other innovative firms. Many variations on this technique have been published and are used. Here, we look at the props, personnel, and rules required for one form of structured brainstorming.

There are a few physical props required for structured brainstorming:

1. Large flip charts
2. Adhesive tape
3. A large felt-tip marker

The flip charts are used to record the proceedings of the brainstorming session; adhesive tape (masking tape works well) is used to secure the flip-chart sheets to the walls in the meeting room; and the felt-tip marker is used to record the proceedings. Sometimes people suggest that a blackboard be used, but this is inferior to flip charts because most rooms do not have enough space to record a full meeting and blackboards are not a permanent record. If a session is going well, dozens of flip charts can be filled in a matter of minutes, and it is very easy to get a meeting record of 25 to 50 sheets. It is also unacceptable to forgo a common public recording of the proceedings; it is important that individuals know and agree to what is being recorded, and it is important for everyone to have access to the "shared memory" of the full meeting record at any time.

Brainstorming works best in groups of three or more, and during the session all members of the group must have equal status, whether they belong to management or to the rank and file. Initially, one person volunteers or is designated to be the scribe. He or she stands at the flip chart and records the proceedings as accurately as possible. The scribe is not a meeting leader and is authorized only to record, not filter, information. He or she is allowed to ask questions to clarify a point so it can be accurately recorded and may call attention to procedural matters, but while a person holds the marker as scribe he or she is prohibited from making a creative or critical contribution to the session. This rule is very important: It prevents the scribe from slipping into a "leader" role by prohibiting him or her from taking too active a part. To be fair, however, a scribe wishing to offer a contribution to the session can become a participant by handing the marker to another group member, who then becomes the new scribe. Scribehood, in this way, can and should be shared by different group members.

With proper props and a scribe in place, the structured brainstorming session can begin. In its normal course it follows six steps in the following order:

1. Discuss the mess.
2. Define the problem.
3. Generate solution alternatives.
4. Generate and select solution criteria.
5. Rate alternatives according to the criteria.
6. Select a proposed solution or set of solutions.

In the remainder of this section, we examine what is meant by each of these steps and why they are so ordered.

Discuss the Mess

Sometimes problems are simple enough that short briefings by a group leader are enough to understand the difficulty. More often than not, however, real-world problems are fairly complex, requiring the input of many group members to flesh out their full extent. During this initial phase of brainstorming, members discuss the mess; that is, they bring up background information, historical information, the present situation, other solutions that have been tried, and any other information that may help the group understand the difficulty. During this phase it is important to require that there be no debate or argument. Group members may independently present their different views of the world and all views are recorded by the scribe. There should be no attempt to organize the material in any way, and members should be encouraged to associate freely. Contributors should be brief and to the point; long speeches and war stories should be discouraged. Additionally, some effort should be made to avoid being prescriptive at this point because there is as yet no problem to solve. There will be ample time for generation of alternatives later in the process; it is more productive during this phase to concentrate on symptoms of difficulty and hypotheses regarding root causes.

After discussing the mess for a time, it is common for members to begin to sense that the group's wheels are spinning. Issues are repeated or packaged in new wording, but little new information is coming out. Such repetition is often a signal that the mess has been sufficiently discussed. At this point, the scribe or any other group member sensing a slowdown can ask whether the group is ready to define the problem. It may be useful at this point to go back over the session history, to see if other ideas are generated associatively. Once the group is ready to move on, it is time to define the problem.

Define the Problem

During the discussion of the mess, many issues, both germane and peripheral, are invoked. During the second phase, it is time to focus on which issues should be tackled during the remainder of the session. In terms of the writing model of Chapter 5, discussing the mess is analogous to directed creation, and defining the problem is analogous to revision. One difference between the writing model and a problem-solving meeting is that the meeting is a group activity, and it is important to come to a succinct statement of the problem that the group can agree to. One way to avoid unnecessary conflict is to tend toward inclusiveness in the problem statement. At this stage, as long as the problem definition is fairly well on target, a somewhat larger definition that encompasses the views of the whole group is better than one that arbitrarily excludes some issues important to a significant minority. Thinking ahead to later stages, it may be possible to find solutions that cover those special concerns without much extra effort or cost; even if it isn't possible, the extra concerns can always be discounted at the later stage of formulating a solution or set of solutions.

During the problem definition stage, discussion is permitted but care should be taken to avoid bickering and needless debate. Once a suitable, succinct problem statement has been created, it is time for generation of alternatives to begin.

Generate Alternatives

Generating alternatives is the most exciting part of structured brainstorming. The rules are simple: No holds are barred, all ideas are welcome, and no criticism of any idea is permitted. Not one word. This rule is the moral equivalent of "not crossing out" in freewriting and directed writing. The scribe simply writes down alternatives as they are generated, posting each new sheet of alternatives on the walls around the room. Group members are allowed—they are encouraged—to bounce ideas off one another to create hybrids or embellishments; ideas should flow freely and associatively from one to another. Again, there should be no effort to make ideas come out in any particular order. Human thought is a messy process, and we should let it be so. There will be plenty of time for the harsh light of reality to shine on silly or infeasible alternatives. In the meantime, every out-landish, wacky idea that gets mentioned increases the chances that some creative, innovative, and perhaps more practical idea might pop into someone's mind.

If the problem is sufficiently difficult and more than one session is required, it is often useful to break alternatives generation into multiple sessions. If group members have a chance to "sleep on it," they often return to the table refreshed with alternatives that would not have occurred to them during a single session.

Generate and Select Criteria

Solutions are only good and bad in relation to criteria that they satisfy or don't, and in this step of the brainstorming session a set of criteria for judging the alternative solutions is generated and selected. During discussion of the mess, many criteria typically get mentioned, so the first part of this phase is to collect candidate criteria by reviewing the "mess." Additional criteria can and should be added to the list. As with the other creative portions of the procedure, this first pass at generating criteria should avoid criticism and be inclusive. Thereafter a second pass should be made to ensure that each included criterion is essential to project success. If controversy arises, it is still best to err on the side of inclusion. In the final debate, spurious criteria will usually be discounted.

Once the list is culled, a decision must be made on how to score each criterion. This can be as simple as a qualitative judgment of effectiveness $(+)$, ineffectiveness $(-)$, or indifference $(\pm)$, or a simple subjective score (say, 1 to 10). Some criteria lend themselves to a more quantitative evaluation of a relevant statistic, such as expected profit, volume, sales, and so on. Once the choices are made, each alternative solution can then be rated according to the list of criteria.

Rate Alternatives According to the Criteria

Once the alternatives are generated and the criteria are chosen, alternatives can be ranked according to each criterion. The easiest way to go about this is to make a matrix with alternatives listed down one side and criteria listed across the top. In the usual alternatives generation session, some of the solutions will be basic configurations, and others will be features or refinements that can be added to (or taken away from) one or more base configurations. For example, in solving the problem, "obtaining personal transportation to and from work," base configurations might be

1. Buy a vehicle.
2. Lease a vehicle.
3. Take the bus.
4. Ride a bicycle.
5. Walk.

Refinements that might naturally arise during the generation of alternatives could include a listing of specifications of the cars that might be purchased or leased, the type of purchase or lease plan, arrangements for maintenance, and so forth.

When listing each feature in the evaluation matrix, it is convenient to group features with their base configurations so they can be considered for inclusion or exclusion independently.

Select a Solution or Set of Solutions

After the matrix is filled out, it is time to do some deciding, or at least some culling of the list. There are formal decision-making procedures for using multicriteria ratings such as these, but often the process of going through the brainstorming exercise and filling out the evaluation matrix will sufficiently focus the group's attention on the best solution or solutions. If this happens, great! If not, it is likely that two or more subgroups feel that there are significantly different solutions that are best for the organization. In these cases, the best thing to do is not to seek compromise. Subgroups espousing different solutions should hammer out separate proposals, and the final decision should be made in a manner consistent with normal organizational decision-making procedures.

Sticker Voting: A Quick-and-Dirty Shortcut

Some decisions deserve the full brainstorming treatment as described above. For others, either the costs of having a group sit around and go through the entire procedure are too great or action is required fairly quickly and cannot wait for the full procedure to run its course. In these cases, there is a useful abbreviated scheme that can quickly determine whether there is consensus on the outline of a solution.

In this shortcut, called sticker voting, the brainstorming process begins with the first four steps: (1) discuss the mess, (2) define the problem, (3) generate alternative solutions, and (4) generate and select solution criteria. (If time is really pressing, even the fourth step can be dropped.) Thereafter members of the group are each given a set of colored stickers and are asked to vote by placing their stickers directly on the flip-chart sheets next to the elements of a solution they favor. A somewhat chaotic scene usually follows, with members placing their stickers, horse-trading votes, and bumping into each other as they make their decisions.

After the dust settles, large clusters of stickers identify the most-favored elements of a solution, and the number of sticker votes may be recorded and passed along as the group's recommendation. The shortcut relies fairly heavily on the group members' intuition regarding the connection between solution elements and criteria; by not going through the formal process of considering solution elements against each of the criteria, it is possible that group members will miss important trade-offs in making their evaluations. Nonetheless, in cases where a quick or inexpensive recommendation is necessary, abbreviated brainstorming with sticker voting brings the benefit of a full discussion of the mess and alternatives without the protracted evaluation of solutions against criteria.

9.5.5 Putting Structured Brainstorming to Work

The process we have examined is fairly straightforward, and if the rules of engagement are followed closely, the result can hardly help but be an improvement over the usual serial idea–criticize–idea–criticize approach adopted in most unstructured meetings. To get some experience using the technique, try the next exercise.

The beauty of structured brainstorming is in its ruthless separation of the creative and critical components of the process, as well as its prevention of the leader–follower group dynamic. Practiced regularly, it can help boost the quantity and quality of the solutions created by any group with which you are associated.

Exploration Exercise

Apply structured brainstorming to the problem of finding engaging employment.

SUMMARY

In this chapter, we have considered the current popularity of forming teams in modern organizations in light of some of the obstacles and opportunities for so doing.

We started by distinguishing between groups and teams by examining recent definitions from current teamwork research. One key difference is that working groups rely on individual contribution and accountability, whereas teams rely on combined effort and mutual accountability. These distinctions led to a discussion of team ground rules and how they help align expectations.

Thereafter, some of the difficulties of forming effective teams were considered. First, the complexity of teams was examined in light of two little models: the deciding–doing model and a model of the probability of team conflict. These led to discussion of the prisoner's dilemma problem recast as the teammate's dilemma problem to help understand why team cooperation is so difficult to achieve. Oftentimes, individuals who cooperate with the team goals are forced to carry freeloaders who choose to do less than their fair share. Understanding this dilemma adds force to the establishment of team ground rules and norms.

Finally, we recognized that teams meet fairly frequently and that even modest improvements in meeting efficiency and decision quality can greatly improve organizational productivity. The simple measures of starting and ending on time and following an agenda were advanced as one way to ensure that meetings are more effective. In addition, a structured methodology of holding brainstorming or problem-solving meetings is an effective means to getting ideas discussed and then assembled to form high-quality solutions.

EXERCISES

1. Consider a group project experience in your past where significant freeloading was in evidence. Write a short essay describing the situation, how the project was completed, and the interpersonal conflicts that took place.

2. Consider the best team or group experience of your life. Write a short essay describing the situation. Identify why you now view the experience as positive. Discuss your role and that of other key members of the team or group in making the experience a positive one. Despite the positive nature of the memories, also describe any freeloading or other nonproductive behavior by team members and its impact on the team experience.

3. Form a group of three or more members, and apply the structured brainstorming procedure (or the abbreviated scheme with sticker voting) to "solve" a current problem in your organization.

4. Form three or more teams consisting of three or more individuals each. Select a current problem facing your organization and have each team solve the same problem using the structured brainstorming protocol. After the solutions are complete, convene a meeting of all teams and compare and contrast the selected solutions.

5. Select a problem that was recently solved in your organization by traditional, unstructured means. Perform structured brainstorming on the same problem, individually or in a group. As much as possible, ignore the previous solution. Compare and contrast the solutions derived by structured and unstructured means.

6. Consider the deciding–doing model of team efficiency [Equation (9.2)] and apply it to a team project you have worked on. Specifically, tally the total time spent in meetings versus the total time spent working on getting the project done. Calculate the optimal team size. Consider whether your team is larger or smaller than that number.

Chapter 10

Organizations and Leadership

10.1 ORGANIZATIONS AND LEADERSHIP MATTER

It is important to cultivate the individual and team skills necessary to be an entrepreneurial engineer, whether those skills are exercised in a startup, as a freelance, or as a member of a larger organization. And those who do start companies or rise to high levels within existing organizations have the special opportunity to design organizations and lead them. But more frequently what one needs to know about organizations and leadership has less to do with designing or running an organization and more to do with identifying (1) the kinds of organizations we wish to work for and (2) the kind of leader we want to follow and become.

These two items require that we understand good organizations and leaders, and we start by asking what principles of human behavior or motivation can help us analyze alternative organizational structures and leadership strategies. This leads us to Maslow's famous hierarchy of needs and McGregor's classic study of theory X and theory Y organizations. These classic works lead us to two current empirical studies—one on organizations and one on leadership—that examine common attributes of success in each domain. The two studies are united by their quest for understanding attributes of excellence, but we also consider a more sociological approach that surveys organizational *culture*, looking for stable regular patterns of organizational life and reasons why those cultural patterns persist. We conclude by asking what appears to be a naïve question about why we join organizations in the first place. To answer the question on economic grounds is surprisingly difficult, and the answer leads us to a fuller understanding of why organizational life in an Internet age is so fluid. This leads to a discussion of the increased tendency toward work as a *free agent* and the need to organize one's work output in a *portfolio,* much as artists have always prepared portfolios of their most important pieces.

10.2 UNDERSTANDING HUMAN BEHAVIOR AND MOTIVATION

At the level of pairs of human beings, a key to good human relations is the golden rule. One on one, this seems sensible, but it is difficult to know how to generalize these lessons to groups or organizations. We start by asking if there are bounds on human behavior that help us delimit what people might do in particular situations. The answer turns out to be negative, and we turn to *human motivation,* Maslow's hierarchy and McGregor's theory X and theory Y for more effective guidance.

10.2.1 Bounds on Human Nature

Looking on the brighter side of human nature, we know that, among other things, people can be

- Gentle
- Contemplative
- Insightful
- Punctual
- Neat
- Polite
- Trustworthy
- Creative
- Energetic
- Consistent

Looking on the darker side, we also know that people can be

- Brutal
- Oblivious
- Uninspired
- Tardy
- Sloppy
- Rude
- Untrustworthy
- Dull
- Lazy
- Inconsistent

Apparently, there are few bounds on human behavior, and, to make matters worse, we can find the best and the worst of these behaviors exhibited by a single individual.

Looking at behavior alone is of little help in our attempt to model groups of humans, but this should come as no surprise because behavior is merely an effect, not a cause. To get at underlying causes we must better understand human motivation.

10.2.2 Unifying Model: Maslow's Hierarchy of Needs

There is a straightforward model of human motivation—Maslow's (1987) hierarchy of needs—that is a favorite of management theorists and practitioners. The model is popular because it is relatively simple to convey and fairly easy to apply in practical situations.

Maslow's model begins with the premise that human beings are needs-driven animals and goes on to say that human needs are organized more or less hierarchically, from basic to more complex (as depicted in Table 10.1). Thus, as a person fulfills basic bodily needs (such as food and oxygen), he or she begins to focus on higher needs (such as safety). As these are increasingly satisfied, needs for social interaction become more important, and so on down the list (which is up the hierarchy).

Models are useful to engineers for their real-world predictive power, which this one seems to possess. Applied to the bewildering array of behaviors surveyed in the last section, it makes it possible, perhaps, to understand what motivates a behavior that we might otherwise be tempted to label "good" or "bad." For example, when an employee is evasive and defensive in the face of repeated questioning by a negative, distrustful boss, it is easy to understand the employee's fear of losing a job and to see how basic needs such as food and security are at stake in the conflict. Likewise, it becomes easier to understand the extra effort exerted by an employee who has been repeatedly praised by a positive manager if we understand the ego gratification such praise provides.

Of course, organizations and their leaders interact strongly with the needs hierarchy of all their individuals, and it is interesting that as we move from more basic concerns (body, safety) to higher needs (social, ego, development) we find fear yielding to happiness as the primary emotion behind the motivation. As long as there have been sticks and carrots, managers have intuitively recognized the fear–happiness dichotomy in their dealings with their people. The dichotomy

Table 10.1 Maslow's Hierarchy of Needs

Needs	Examples
Body	Oxygen, food, procreation
Safety	Industry safety, home security
Social	Sense of belonging, friendship
Ego	Recognition, praise
Development	Aspirations, striving for excellence

has also been explicitly recognized in the management literature, and in the next section we'll consider two theories of McGregor and their implications for modern organizations and their leaders.

10.2.3   Theory X and Theory Y

McGregor's (1985) seminal book, *The Human Side of Enterprise*, considers the management of organizations from the standpoint of human motivation, using Maslow's hierarchy as an important base model. He concludes that there are two fundamentally different types of managers who operate from two different sets of assumptions about human behavior.

The first type of manager operates from a set of assumptions that McGregor labels *theory X*. These managers believe that people:

1. Dislike work.

2. Are not trustworthy and need to be watched constantly.

3. Cannot make good decisions without close supervision.

4. Need prodding to complete even the simplest of tasks.

A manager believing these things is naturally led to a suspicious, coercive management style, where careful monitoring and punishment are the rule. By contrast, what McGregor calls the *theory Y* manager believes that people:

1. Find work enjoyable.

2. Can be trusted to work even in the absence of supervision.

3. Can make good decisions autonomously.

4. Can finish tasks without constant prodding.

Because of these beliefs, the theory Y manager is led to a more supportive, praise-oriented style of management.

It seems odd that two such disparate viewpoints can coexist. After all, managers aren't generally stupid, and they are likely to adapt over time in trying to improve. How could such different views coexist in the long run? One answer is that both strategies are *locally optimal*. A solution is "locally optimal" if small changes in decision variables about the current operating condition result in a degradation in performance; thus both theory X and theory Y managers find that there is almost no way to improve their performance in the neighborhood of the chosen solution.

For example, as theory X managers flog their people—as they exert their suspicious, manipulative, often punishing brand of management—organizational output improves to a point; the people under their control tend to react defensively, reluctantly, and sometimes suspiciously, thereby confirming the theory X viewpoint. On the other hand, theory Y managers, by trusting their people and delegating authority and responsibility, find increased output; thus their worldview is also confirmed. This seems strange, that both types of manager are locally correct, but the nonlinearity of human behavior—the nonlinearity that is implicit in Maslow's hierarchy—is enough to permit multiple optima to exist.

At this point, we might be tempted to say that we simply have two different solutions, each with its own merits. But, as McGregor goes on to argue, theory Y is the better approach in the long run because it appeals to higher motives by emphasizing ego gratification. This is significant, as once employees take responsibility for their lives and seek goals autonomously that enrich individuals and the organization together, the theory Y manager has an easier time keeping the enterprise running smoothly.

Contrast this to the situation faced by theory X managers. The need to keep things running requires increasing threats to maintain the same, if not diminished, levels of performance. Monitoring efforts are effective as long as theory X managers are standing there; as soon as they turn their backs, the troops look for ways to slow things down, gum up the works, or otherwise obstruct progress. Over time, the theory X coercive style requires an escalation of threats that are decreasingly effective.

Almost any organization of any size is filled with its own tales of theory X and theory Y. The next section considers one true story in which an employee's behavior transformed dramatically with a change in management.

10.2.4   The Case of the Sluggish Secretary

Consider the case of the sluggish secretary. There was once a secretary in a government agency who was reluctant to do any work not part of her normal routine of phone reception, filing, and light typing. Department co-workers who tried to approach her with out-of-the-ordinary jobs were met with evasion and avoidance, and many shared the view that she was not helpful and less than competent.

At the time, the secretary worked for a boss who was quick to find fault and slow to forgive. One day, job assignments were switched, and she was assigned to work with an up-and-coming young manager. He, too, had had his share of bad experiences with her, but he was too busy to dwell on the past and immediately gave her responsibility to organize a seminar series. The job required that she perform many new tasks, including letter writing, mass mailing, phone calls, sales calls, and follow-up, as well as on-site registration and client contact. To everyone's surprise, this once-sluggish secretary attacked her new responsibilities with a vengeance, the seminar series was a success, and the woman's confidence and helpfulness spilled over into her other duties. When asked to do new things, she no longer avoided or evaded them. She simply tried to dig in and do the job.

In looking back over this description, we can easily see what accounts for the change. She went from a theory X situation, with suspicion, punishment, and blame lurking at every turn, to a theory Y situation, with responsibility, trust, reward, and responsibility the watchwords of the day. The transformation that occurred can be cited repeatedly in stories of rapid change in outlook, attitude, and performance. The magic that theory Y worked in this case is not unusual, and the exercise below asks you to consider whether these lessons apply in an organization with which you are familiar. For the reader interested in other war stories of theory Y in practice, see Robert Townsend's delightful book, *Further up the Organization* (Townsend, 1984).

Exploration Exercise

Consider an organization with which you are familiar. In a paragraph or two, explore whether it is a theory X or theory Y organization and explain your reasoning.

10.3   HUMAN ORGANIZATIONS AND THEIR LEADERS

Organizational theorists have contemplated the tie between human motivation and organizational design since the time of Maslow and McGregor, but modern organizational theory and empirical research are confirming the tie between sustainable organizational excellence and those organizations and leaders that bring out the best in their people. These studies give important clues as to the kinds of organizations and leaders to work for. The studies also provide role models for the engineer to emulate when he or she is charged with leading a team, department, or organization.

Here we briefly consider two studies, one that examines companies with a track record of sustained high performance in an attempt to understand the common attributes of their success and a second that examines top leaders.

10.3.1   From Good to Great

What separates merely good companies from great companies? That was the question asked by author Jim Collins in his book *Good to Great* (Collins, 2001). The book reports the results of a 5-year study seeking common features of companies that transitioned from average performance or worse to sustained excellent performance over at least 15 years that was three times that of the market. This was a difficult filter once we recognize that General Electric only (only?!) beat the market by a factor of 2.8 from 1985 to 2000.

After running the numbers, 11 companies from the Fortune 500 made the cut, and they are presented in Table 10.2. Perhaps somewhat surprisingly, the company list is nowhere near as glamorous as conventional wisdom might expect. Banks, retailers, steel, and cigarette companies are not the high-tech poster children that some might have expected. After selecting the companies, Collins and his research team analyzed each of the selected companies against a corresponding comparison company in the same industry. They also considered six companies (Burroughs, Chrysler, Harris, Hasbro, Rubbermaid, and Teledyne) that made the shift from good to great but did not sustain greatness for the requisite 15 years.

Table 10.2 Good-to-Great Companies (Collins, 2001)

Company	Market Factor
Abbott	3.98
Circuit City	18.50
Fannie Mae	7.56
Gillette	7.39
Kimberly-Clark	3.42
Kroger	4.17
Nucor	5.16
Philip Morris	7.06
Pitney Bowes	7.16
Walgreens	7.34
Wells Fargo	3.99

The data collection effort for the study was elaborate. The project systematically coded 6000 articles on the subject companies and consumed 10.5 person-years of effort. Interestingly, the project did not start with explicit theoretical assumptions. It sought to mine a model from the financial, company, and business press data they analyzed. Out of this 5-year effort, Collins presents eight common threads among the good-to-great companies. Here we consider four of the key items as follows (Collins, 2001):

1. Humility + will = leadership.
2. First who ... then what.
3. Confront the brutal facts (yet never lose faith).
4. Simplicity within the three circles.

Each of these features of good-to-great (GTG) companies is briefly examined. For a more detailed discussion, consult the original text.

Humility + Will = Leadership

Americans love the charismatic leader who charms his way to success. No one around at the time will forget Lee Iacocca leading Chrysler with television commercials that intoned, "lead, follow, or get out of the way." But sustained good-to-greatness was not achieved by that kind of leader. Instead, Collins observes that sustained turnaround is led by those who combine *humility* and *will*. Humility counters our usual notions of leadership, but the study found numerous instances of egotistical leaders getting in the way of finding and executing steps necessary for success. Will is not often associated naturally with humility, but the good-to-great leaders were humble at the same time they knew exactly where to focus the energies of themselves and their people.

First Who . . . Then What

Many modern books on organizations start with two "ions"—miss*ion* and vis*ion*—but good-to-great leaders built great companies by selecting the right people and helping them find a way to greatness. The right kind of people do not need to be motivated. They are self-motivated, and they need very little management.

Confront the Brutal Facts (Yet Never Lose Faith)

When faced with bad news, many organizations resort to excuse making and refuse to look bad news in the eye. Good-to-great companies refuse to lie to themselves. They seek solutions to tough problems through questioning, dialog, and investigations without blame. In this way, GTG companies rapidly find and fix problems and move on to opportunities.

Simplicity within the Three Circles

Sticking to your competencies is a modern management mantra, but good-to-great companies deeply understand the business they are in and have figured out how to reduce sustained profitability to simple strategic and financial formulas. In doing so, they understand (1) what they can be the best at, (2) what is their economic "denominator," and (3) what is their passion.

The first of these is the usual notion of business strategy—to understand one's niche—but GTG companies were able to identify, simplify, and focus on a single item relentlessly.

Second, all organizations have metrics they watch, but GTG companies try very hard to maximize their profit per unit "something," where "something" is chosen with great care. For example, Walgreen's shifted from a focus on profit per store to profit per customer visit, thereby tying repeat customer satisfaction and location profitability together in a single number.

Finally, GTG companies understand that managers and workers aren't interchangeable cogs in corporate machines. Human organizations run on passion for creating great products and services, but for much of the 20th century, business schools have emphasized improvement within functional specialties such as finance, marketing, accounting, and strategy as the way to excellence. Understanding best practices in these different disciplines is important, but such understanding is no substitute for being fundamentally committed and passionate about the products and services one helps to provide. In the old days, it was easier for graduating engineering students. Car guys went to car companies, gizmos went to electrical engineering firms, and mud-in-the-boots civil engineers tromped around some construction site for a living, and the managers these people worked for also came up from the trenches covered with grease, solder, or mud. Today's engineers need to think about their passions and the passions of the people they work with. Understanding and following your passion, and recognizing the passion (or lack) of the people you work with and for can be critical clues in assessing your affiliation with an organization.

Exploration Exercise

Consider an organization with which you are familiar. In a brief essay, identify the ways in which the organization does or does not follow the principles of a good-to-great organization as discussed above.

10.3.2 The Leadership Challenge

Understanding great organizations and organizational culture are helpful in evaluating whether you might want to join a particular organization and how to improve the organization you belong to. Knowing which leaders to work for and how to improve your own leadership skills are equally important. Much has been written about leadership, and the subject is vast and complex, but a contemporary empirical study identifies important dimensions of leadership that are quite helpful in evaluating leadership in yourself and others. In particular, Kouzes and Posner originally published *The Leadership Challenge* in 1987 (Kouzes & Posner, 2003) based on work begun in 1983. Ordinary middle- and senior-level managers in many public- and private-sector organizations were asked to describe extraordinary experiences of leadership. Specifically, surveys were conducted with 38 open-ended questions asking the subjects to reflect on their experience, its context, their preparation, special techniques involved, and so forth. The original book was based on 550 long surveys and another 780 short-form (2-page) surveys together with 42 in-depth interviews. The research yielded a model of leadership as well as a quantitative instrument, the Leadership Practices Inventory—and the model and the instrument have been validated by subsequent testing on over 10,000 leaders and 50,000 other constituents. Interestingly, the work refutes many of the common stereotypes of leadership, but it does suggest that leaders exhibit certain common practices at their best.

In particular, Kouzes and Posner (2003) suggest that there are five fundamental practices of exemplary leadership:

1. Challenge the process.
2. Inspire a shared vision.
3. Enable others to act.
4. Model the way.
5. Encourage the heart.

Each of these is briefly discussed in what follows.

Challenge the Process   Many of the leaders in the Kouzes and Posner study (2003) attributed their successes to "luck," but most of them were activists in setting up the favorable circumstances that led to their good fortune. All cases of personal-best leadership involve some kind of challenge, and all cases involved a change to the status quo. Thus, leaders must (1) seek new opportunities and (2) foster an environment conducive to risk taking and experimentation. Many organizations pay lip service to these values, but putting them to practice requires a special attitude toward innovation, a healthy skepticism toward comfortable routine, and a rare tolerance of learning from mistakes.

Inspire a Shared Vision   Upon challenging the process, many effective leaders understand that they need to communicate a *vision* of the future to inspire others to change. This requires (1) understanding the past, and (2) imagining the ideal, (3) outlining a viable future with unique business opportunities, and (4) enlisting others to the cause.

Enable Others to Act   Once a new direction is set, it is important to encourage action by others, and this requires collaboration and information. Sustainable collaboration requires a win–win exchange between the parties involved. In turn this requires an analysis of what you want others to do and what they wish of you. Posner and Kouzes (2003) suggest an analysis of the "currencies" or media of exchange between parties to encourage the needed collaboration.

They also recommend the sharing of important information. Many managers reflexively hoard information like squirrels hoard acorns. Informed action by others requires sharing of key pieces of strategic information. An extreme version of this sort of sharing is found in Stack's style of *open-book management* (Stack, 1992), in which the financial books of a company are opened to its employees. Not every business may be comfortable with this level of sharing, but the spirit of such openness was an element in best leadership practices as reported by Kouzes and Posner (2003).

Model the Way   "Do as I say, not as I do," could be the watchword of management in the scandal-plagued early years of the 21st century, but personal-best leadership required that leaders "walked the talk." This observation is consistent with the good-to-great study (Collins, 2001) discussed elsewhere. Chicanery and boosterism can work in the short term, but long-term achievement requires a steady state of kept promises up and down an organization. The positive attitudes of a theory Y organization and manager are entirely consistent with the empirical results of the Kouzes and Posner (2003) study.

Encourage the Heart   Organizations that try to buy employees can always lose out to a higher bidder, but organizations that evoke strong positive emotions are much harder to leave. This is not to say that pay and perquisites are unimportant, and the evidence does suggest that personal-best leadership requires a match between those variables and performance. But many instances cited by Kouzes and Posner (2003) involved organization members going above and beyond the call of duty. In many of those cases, the only reward for doing well was acknowledgment, an attaboy, a thank-you note, or public recognition of accomplishment. As has been discussed elsewhere, praise is a powerful force, but too often it is handed out sparingly, if at all.

10.4 ORGANIZATIONAL CULTURE: THE GODS OF MANAGEMENT

The *Good to Great* (Collins, 2001) and *Leadership Challenge* (Kouzes & Posner, 2003) studies examine organizations from the standpoint of *best practice*. This is an effective methodology for mining data for common attributes of success, but it is less helpful in understanding ordinary practice and the typical organizational patterns we might observe in a larger population of leaders and organizations. Here, we take a more sociological viewpoint and try to identify different organizational and leadership through an understanding of *organizational cultures.* We try to understand how culture evolves in response to the type of activity the organization pursues.

A useful guide to understanding organizational culture was written by Handy in his book *Gods of Management* (Handy, 1995). Handy identifies four primary types of culture and uses Greek gods to symbolize each culture as shown in Table 10.3.

Club Culture

Startups, small family businesses, and other organizations that take their lead from a single talented leader often evolve toward a *club* or *Zeus culture.* In the Greek pantheon, Zeus was loved, feared, and obeyed, and the Zeus culture loves, fears, and obeys its leader. Modern management practice sometimes sneers at such a setup, but there is a clarity and lack of ambiguity in such an organization. My first job out of school was in a small, engineering software firm run by a talented entrepreneur I will call Mack. Whenever I had any question of what I should do, I simply asked myself what Mack would do, and I did it. In this way, club cultures communicate through a mechanism of *subordinate empathy*, and in so doing, the culture is able to move rapidly in business climates where speed of decision is

Table 10.3 Four Gods of Management

Culture	Greek God
Club	Zeus
Role	Apollo
Task	Athena
Existential	Dionysus

critically important. Clubs can be stifling for those of independent mind, and in my own experience, after about 4 years of thinking like Mack, I was ready to think for myself, and returned to graduate school. Yet, the experience was deeply satisfying at the time, and joining a club culture run by a talented individual early in your career is a useful way to model successful business behavior.

Role Culture

Apollo was a god of order and rules, and Apollo is the god of the *role* or *bureaucratic* organization. Bureaucracies arise in response to *stable, predictable* environments, and different individuals within the organization take on different roles that are competently (hopefully) executed repeatedly into the future. Government agencies and large corporations in regulated or monopolistic industries often evolve toward Apollo. Although we don't usually think of bureaucracies as efficient, in a stable environment, Apollo effectively and consistently gets the job done. On the other hand, Apollo dislikes and reacts badly to change, regardless of how much it is needed. During times of competitive pressure or an otherwise unstable business picture, the role culture can be inappropriate to meeting the challenge.

Task Culture

The warrior goddess Athena is the symbolic god of the *project* or *task* culture because of her problem-solving talent. In many engineering activities (e.g., design, construction, and consulting), organization success is closely identified with the effective solution of a sequence of discrete problems. Successful project or task cultures allocate time, personnel, and money to the solution of problems, but projects by their nature tend to require customized solutions; they resist solution by routine or rote. Thus, Athena requires innovative and creative individuals who can get a project done well and within time constraints. Because of the customized nature of the solution, and because innovation and creativity tend to be open-ended activities, Athena cultures are expensive. Nonetheless, engineering professionals oftentimes find themselves inhabiting task or project cultures, and the creativity required of the engineering professional is generally a good long-term match to such a culture.

Existential Culture

Because Dionysus is the god of wine, song, and living life to the fullest, Handy chooses him as the symbol of the fourth and final culture. The *existential* culture takes its name from the branch of philosophy called *existentialism*. Existentialism explores the consequences of life as a set of co-existing individuals, largely unconnected from others or without some higher purpose. In many organizations, the individual is subordinate to the larger needs of the organization, but in an existential culture, the organization exists largely to serve the needs of the individuals in it. Individuals with unique expertise such as medical doctors, lawyers, and university professors inhabit such cultures, and they have a large degree of control over the terms of their employment, their bosses, and their customers. In universities, the job of being a department head is likened to herding cats, and the description is apt. It is very difficult to get doctors, lawyers, or college professors to do things they don't want to do. Interestingly, universities, hospitals, and large law firms have parallel universes containing other employees with talents that are more fungible than those of the principals. Those portions of the organization tend toward task or role structures, depending upon their function, but the primary actors tend to go their own way. Existential cultures place a premium on unique expertise, and they do what they need to do to cater to the needs of their special inhabitants.

Cultural Impedance

One might be tempted to ask which culture is best, but the question would be misplaced. As was discussed, different cultures are more or less effective depending on the nature of the work to be done. When speed is of the essence, the club culture can move quickly with one mind. When environments are stable, role cultures can push the paper and get routine jobs done routinely. When problems need to be solved, task cultures solve them, albeit at considerable expense. When unique expertise must be nurtured and cultivated, existential cultures coddle the experts. Which type of organization you might join depends on the level of autonomy you seek, the kind of work you want to do, and the type of expertise you possess. Moreover, it is important to recognize that most large organizations are rarely of one pure type. Different departments within a larger organization or even different teams within the same department will naturally adopt cultures appropriate to their mission.

In thinking about culture, one thing to consider is whether the culture of the organization is appropriate to its mission or whether there is a *cultural impedance* mismatch between mission and culture that is preventing the organization from reaching higher levels of performance. Impedance mismatches will most often be noticed around times of transition. A role culture facing deregulation, for example, will have difficulty adapting to the changing times, or a project culture facing cost pressure may have difficulty standardizing enough to be cost competitive. Such transitional times can be unnerving and difficult, but the greatest opportunities for having a large impact on an organization can exist at exactly such times. Whether one chooses to be a part of an organization in transition or not, recognizing mismatches of cultural impedance is an important skill if one is to evaluate and react to such situations.

Exploration Exercise

Identify four organizations or companies, one for each of the four types of organizational culture: club, role, task, and existential. In four short paragraphs, consider the ways in which the culture aids and hinders each organization in achieving its mission.

10.5 WHY FORM OR JOIN ORGANIZATIONS?

Much of this chapter has assumed that it is quite natural for us to form and join organizations as a regular feature of modern work life, but some strange things are taking place in the age of the Internet. Organizations appear to be getting smaller; functions that were once a regular part of an organization are being outsourced, and more and more people are working as temps and freelancers. To better understand why this is happening and whether there is a connection between the Internet and these workplace tendencies toward leaner, more fluid organizational structures, we start with a naïve, yet fundamental question. Why, from an economic perspective, do we form, and join, organizations? The answer is not obvious, but it is important and helps explain the tendency toward leaner organizations with specialized functions than in times past.

At first, the question sounds almost silly. After all, forming organizations is what you do when you want to get organized, right? Yet the questions of appropriate leadership and organization only matter if we *choose* to join, and behind the question is a serious matter. Since the fall of the Soviet Union, it seems that much of the developed world now agrees that free markets are a sensible way to organize economies (and democratic rule is a sensible way to organize political systems, but that is a question for another book). One scholar has gone so far as to suggest that the widespread acceptance of free markets and democracy signals an end to socioeconomic evolution, the end of history (Fukuyama, 1992). Whether this is so is beyond the scope of this argument, but this end of history leaves us with a puzzle that is germane to our investigation of organizations and leadership. If free markets are such a great thing, why do most of us as individuals choose to *suspend* the free market in our choice of employment? The question sounds strange, but think about it. Every morning we arise and we face a choice. We could choose to get up and sell our services to the highest bidder, but instead most of us get up and simply go to work under an arrangement that was negotiated at the outset, but no longer is directly subject to market exchange.

Questioning the common sense of getting up and going to work is odd, but we must do so if we are to get at the economic forces that cause us to do this. After all, much of what happens in business is shaped by economics, and we must understand the fundamental economic forces that result in the formation of organizations if we are to put our query onto solid ground. To help make what might be an abstract discussion more concrete, imagine two structures with identical numbers of workers, one normal, fairly stable, and one where the free market is consulted on a regular (monthly, weekly, daily, or even hourly) basis.

Consider a small manufacturing *organization* with a boss-entrepreneur and a single employee (Figure 10.1). Early in the company's history, the entrepreneur designed the widget, started the company, lined up clients, set up the initial manufacturing process, and performed the initial production runs. After a time, the boss gets tired of being a one-man band and hires an employee. The boss still

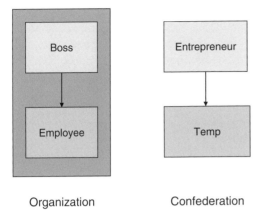

Organization   Confederation

Figure 10.1 Schematic illustrates how subtle the difference is between an organization and the looser confederation.

gets the contracts, designs the widgets and manufacturing equipment, and does the accounting. The employee now works the widget assembly equipment, packages the widgets, and ships them to the customers. All this sounds fairly normal, but the simple example is enough to help make our point.

Consider an analogous situation that starts with the same boss-entrepreneur, except this time when the boss gets tired of working alone, he is aided by a sequence of *temporary employees* (temps) who are hired by appealing to the market with high frequency. To give this combination a convenient name, we will call the combination of the entrepreneur and the temp a *confederation* because the temp agrees to work for the entrepreneur for a brief and predetermined period of time. (See Figure 10.1.)

Although temporary employment has become more common, it is still rare that any of us will take a job on a daily or an hourly basis. Why are we so willing to suspend the free market and join organizations? Without the careful setup, we might have been tempted to answer with Adam Smith's famous conception of *division of labor* (Smith, 1776/1937). Indeed, in going from a one-person organization to a two-person organization, the entrepreneur understands the value of *specialization* to economic productivity, but note one crucial thing. The division of labor is identical in both the organization and the confederation. Therefore, division of labor cannot help us explain why we are so promiscuous in forming and joining organizations.

This is something of a puzzle that grows worse as we continue to ponder the two structures. There just don't appear to be many significant differences between them at first. Both structures make the same widgets; both structures have the same number of people in the same roles. The *only* difference between them is that the organization chooses not to turn to the free market as frequently as does the confederation to hire its help (nor does the employee choose to turn to the free market to look for work as often as does the temp). In other words, employers and employees in organizations agree, somewhat implicitly, to form a longer-term union. Recognition of this mutual commitment is very important and should be kept in mind in the study of organizations and leadership of organizations, but to better understand where the basic economic impulse for this comes from, we need to ask what economic advantages organizations have over confederations. Such advantages must exist, or otherwise entrepreneurs would prefer to turn to the free market to engage helpers for their enterprises.

It is tempting to explain the advantages of an organization directly. For example, perhaps the entrepreneur is motivated to hire a permanent employee—instead of turning to the marketplace to hire a subcontractor—by (1) cheaper wages, (2) better control, or (3) clearer brand identity. However, none of these reasons withstands much scrutiny. If the wages paid to a subcontractor are greater than those paid to an employee, they are greater because of the greater risk taken by a subcontractor facing uncertain duration of employment; in the long run the free market can be expected to evaluate this trade-off between risk and payment accurately. Control is no better an explanation of why organizations exist because the entrepreneur exercises equally clear management authority in both an organization and in our hypothetical confederation. Moreover, because the identity of an organization—or a confederation—is based on the cost, reputation, and quality of what is produced, and because neither entity has an inherent advantage along these dimensions, identity cannot be used to explain why organizations are formed and joined.

If such direct effects fail to explain an organization's advantage, what does? Turning to the essential difference between the confederation and the organization—the confederation's more frequent use of the free market to engage services of individuals—helps shed some light on the organization's advantage. If an entrepreneur turns to the free market to hire subcontractors, chances are that different bidders will be engaged on different days. As a result, the confederation will be subject to a number of costs that a corresponding organization with its more-or-less permanent workforce need not pay:

1. Cost of hiring, paperwork, and contracting

2. Cost of training

3. Cost of incompetent hires

Frequent turnover of subcontractors in the confederation will require that whatever paperwork, legal fees, and other sign-up costs are necessary will be incurred each time a different subcontractor is engaged. Worse than this, each time a new subcontractor is hired, he or she must be made familiar with the local particulars of making widgets, and in most manufacturing processes training costs can be significant. Finally, when one turns to the marketplace, one can never be certain that a hire is going to work out. There are plenty of smooth talkers with good-looking vitas who turn out not to be very good widget makers. This analysis is presented from the entrepreneur's point of view, but similar conclusions can be reached if it is approached from the employee's or subcontractor's perspective.

Taken together, these seemingly secondary factors, what economists call *transaction costs*, are the primary economic reason why we form and join organizations (Coase, 1990). Although some of these costs are mundane things like paperwork and legal fees, many of them are costs of *knowledge* (Sowell, 1996). Confidence in the competence of an employee (or an employer) does not come free of charge; it is knowledge that is gained over a period of time. As a result, when employers find employees they are happy with, and when employees find work situations they are happy with, they tend to stick with each other. In other words, they form an organization, and in so doing they make an implied commitment to one another. It is not a commitment for life necessarily, but it is a commitment not to consult the free market as often as might happen if agreements and knowledge came more cheaply than they do.

Note that this temporal commitment of the individual to the organization and vice versa is not necessarily some warm and fuzzy joining of hands. The search costs for better jobs or employees are simply too high to recommend more frequent attempts to find better ones. In other words, the initial time commitment comes at the behest of economics, pure and simple. There the matter can stand without going a step further, but in many situations the time breeds trust and mutual reliance, and it is this opportunity that is exploited by high-performance organizations.

10.5.1 Optimizing Transactions: A Quantitative Model

Transaction costs in employment are apparently balanced against the difficulty of finding a sufficiently good match between employer and employed. Here we build a simple quantitative model using straightforward engineering reasoning.

Imagine that every time the widget entrepreneur hires a new employee, a transaction cost C is incurred. The hiring of a successful employee results in the employer realizing a marginal value V per time unit, but the probability of success in hiring is assumed to be probability p on each hire. Thus the expected profit P per time period to the employer may be written as follows:

$$P = V[1 - (1 - p)^n] - Cn \qquad (10.1)$$

where n is the number of different employees tried. The optimal number of hires n^* may be found by taking the derivative of profit with respect to n and setting the result to zero:

$$n^* = \frac{\ln C - \ln V - \ln(-\ln q)}{\ln q} \qquad (10.2)$$

where n^* is the optimal hiring frequency (number of employees per period), and $q = 1 - p$. Figure 10.2 shows a particular example of the transaction cost model in which $V = 10$, $C = 0.1$, and $p = 0.1$. According to the formula, the maximum of the curve occurs at $n = 22.4$, and good agreement between formula and curve is obtained. Relating the formula to realistic situations, costs, and probabilities

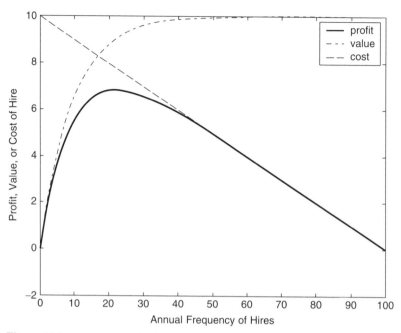

Figure 10.2   Graph of the simple transaction cost model shows a maximum at a particular frequency of hires.

is left as an exercise to the reader; however, the main points at this juncture are to understand that using the free market is not free and that transaction costs act as a kind of organizational glue that bring employers and employees together on economic grounds. Whether or not the economics is a good match to trust or other emotional variables is another question that we must consider.

Exploration Exercise

Calculate the optimal hiring frequency for the situation of Figure 10.1 when $(V, C, p) = (20, 0.1, 0.1)$, $(10, 0.2, 0.1)$, and $(10, 0.1, 0.2.)$. In each case, consider whether n^* increases or decreases. Do these quantitative results match the qualitative discussion of the section regarding transaction costs and the formation of organizations versus confederations? In other words, is organization stability increased or decreased as the value, transaction cost, and success probability increase and decrease?

10.5.2   An Aside on Free Agency

In a globally connected world where résumés, letters of reference, and samples of work can be sent to a potential employer at the click of a mouse, it is clear that transaction costs of employment have reduced significantly. This economic reality has already changed the face of work life as a number of observers have noted (Handy, 1998; Pink, 2001). Handy suggests that the modern global economy is one where the *core* of more or less permanently employed persons is permanently reduced, better paid, and more productive. He captures this idea in the shorthand equation $\frac{1}{2} \times 2 \times 3$—one-half as many people being paid twice as much money doing three times as much work. Pink similarly sees the reduction in permanent work and argues that it is leading—has led—to a kind of free agency in the workforce, where what services you can provide, your network of professional and marketing contacts, and your *portfolio* (Handy's term) of past successes is more important than a job. How far these trends will go before stabilizing is unclear, but there is little doubt that the workplace of today is substantially different than it was even a few short years ago.

Engineering students leaving school decades ago had the expectation of a long career with one or a small number of employers. Today's engineering graduate faces a very different scenario. Unfortunately, government regulations, benefits, and hiring practices remain largely geared to the situation that used to exist. As a result, today's graduating engineer must plan for the world as it is today and make decisions appropriate to employment in such age. To thrive in today's environment, it is particularly important to consider:

1. Portability
2. Loyalty
3. Personal brand and portfolio

Each of these is briefly discussed.

Portability   In a world with shorter terms in the core, longer periods of temporary work or free agency, and ever-changing work relations, benefits tied to a particular employer with long vesting periods and promises in a distant future make much less sense than they once did. Although many company benefit plans still try to secure employees for the long haul, these same companies are less willing or able to follow through on the implied commitment. As a result, other things being equal, insurance benefits that can transfer, retirement and stock plans with short vesting periods, and compensation now versus promises of future increases are to be preferred over traditional schemes that assume a lifetime of work with a single employer. Benefits packages and terms of employment are changing, but the young engineer would do well to seek insurance plans, stock options, retirement benefits, and compensation that can move when the employee chooses or is forced to do so.

Loyalty   Loyalty is dead or so some modern observers would have us believe, but careful examination shows that this is not the case. Yes, given that employment for a single or small number of employers has become less frequent, loyalty of organization to employee and employee to organization has lessened. Yet, loyalty is a natural part of the human condition, and Pink (2001) has suggested that the locus of loyalty is shifting from the organization to one's *network* of support. In other words, the relatively invariant locus of loyalty is not within any particular organization, but instead it is among those co-workers, clients, contacts, or other members of the supply chain of a continuing series of jobs, assignments, or consulting engagements. Interestingly, in the old-style organization, loyalty was also a web of human relationships, but workers tended to identify that web of human relationships with the company itself. From this point of view, rapidly changing times have simply helped us understand that loyalty has always been based in a web of human relationships, not some legal structure, brand name, or organizational entity.

Personal Brand and Portfolio   In times past, keeping an eye open for future job opportunity largely meant keeping you're résumé updated, and back then, when opportunity knocked, it came in the form of a promotion with your current employer or a job offer from a large company in a similar line of work. In a world of limited work at the core and increased free agency, the wise entrepreneurial engineer needs to take more proactive steps to develop his or her *personal brand,* and a key tool toward doing this is the maintenance of a professional *portfolio.*

Modern marketing thinks largely in terms of developing *brand names* for related lines of products and services, and doing so requires a balance of competitive product features, aggressive promotion and advertising, and making consumers aware of the brand, its features, benefits, and consequences. In the same way, work life in more dynamic times demands that professionals think of *themselves* as a product or service; career management transforms to a kind of personal brand management. Free agents understand this because a continued stream of engagements depends on selling themselves to clients, delivering on promises made, and seeking continued or other engagements. The Web and e-mail can be useful tools in promoting your services to others. In addition, maintaining a *portfolio* of successful work product (designs, reports, computer code, etc.) can also be persuasive to potential employers or clients. Engineering students sometimes complain that they don't have practical work experience and thus lack a portfolio. This argues in favor of getting such experience during the summer or in a coop program, but it also argues for looking at one's schoolwork as a source of portfolio items. Excellent class projects, lab write-ups, or capstone design project reports can all be used to start your portfolio to demonstrate your "brand" of engineering service excellence.

Exploration Exercise

Consider yourself as a product. Make a list of your product features, feature benefits, and customer consequences for those benefits. If you were to create a company around your talents, what would you call it? Describe or draw your company's logo. What steps would you take to promote your company to others?

SUMMARY

It makes sense to seek association with the best organizations and to work for the best leaders, but what makes a good organization good and a fine leader fine? To answer the question, "What makes organizations tick?" requires an answer to the question, "What makes people tick?" The chapter briefly surveyed Maslow's model of human motivation—his hierarchy of needs—and continued with a short discussion of McGregor's theory X and theory Y. Essentially, Maslow argued that humans have a complex of needs, and, depending upon their environmental circumstances, they may be motivated by base needs such as survival or by higher order needs such as aspirations toward a more fulfilling life. McGregor took Maslow's theory and applied it to organizational settings by suggesting that some organizations assume that humans are driven by base needs and must therefore be encouraged with carrots (and beaten with sticks) to achieve the desired results. These organizations and managers he called theory X. He also suggested that organizations that believe that high-performing humans are more driven by higher order needs such as personal or societal fulfillment are more likely to respond to a management philosophy built on trust and empowerment (theory Y). Since these early theories, many management theorists have argued that effective management is about getting larger numbers of members of an organization to strive for higher order motivational states.

The chapter continued by considering two empirical studies, one of high-performance organizations and one of high-performance leaders Collins's (2001) *Good to Great* identified attributes of successful organizations by contrasting them to more ordinary ones. Kouzes and Posner's (2003) study presented in their book, *The Leadership Challenge,* identified a number of shared characteristics of high-performance leaders. Although the methodologies of *Good to Great* and *The Leadership Challenge* differ, they fundamentally ratify the Maslow–McGregor view of organization and leadership. Empirically, the best, sustainable organizations and leaders think about their people and how to get them to live fulfilling work lives through growth on the job.

Although focusing on excellence is helpful to understanding improvement, it is also important to understand the more ordinary, everyday organization. To that end, the chapter briefly reviewed Charles Handy's (1995) text, *Gods of Management,* for a useful taxonomy of corporate culture. Handy's four cultures, the club, role, task, and existential culture are all around us, but none of them is inherently good or bad. Handy's point is that culture is *appropriate* (or not) to the business climate faced by an organization. For the engineer, a more important point is whether the culture of an organization is appropriate to the individual's goals and aspirations. Either way, understanding organizational culture and being able to match culture to task and to people is an important skill.

The chapter concluded with an unusual query: Why do we join organizations? Our opening attempt to answer this question suggested that there are economic reasons why we temporarily suspend the free market and take a job. Essentially, using the free market is not free—there are *transaction costs*—and the costs associated with moving from one engagement to another are substantial and cause us to stay put for a time.

The discussion of transaction costs naturally led to the current workplace and the increase in free agency and reduction in lifetime work. As transaction costs of seeking employment and employees have reduced, work has become a more transient affair. These realities help emphasize the importance of portable benefits, loyalty to one's network of contacts, and thinking of one's brand through the maintenance of a portfolio of work output and tangible accomplishments.

EXERCISES

1. Consider an organization in which you have worked. Make a list of the ways in which the organization helped or hindered you in the performance of your job.

2. Consider a manager you have worked for. Make a list of things he or she did that you would classify as theory X behavior, and make another list of things he or she did that you would classify as theory Y behavior.

3. Consider a leader of historical renown. Write a brief essay regarding his or her accomplishments in the light of theory X and theory Y.

4. Scan a newspaper for evidence of corporate or organizational difficulty. Write a short essay analyzing the situation in the light of theories X and Y. Read between the lines when necessary to make your analysis, but be sure to separate fact from speculation in your writing.

5. Consider a corporate leader in the news and evaluate it in the light of Kouzes and Posner's (2003) dimensions of leadership excellence. Write an essay and determine those elements that appear consonant with the dimensions of Kouzes and Posner and those that do not. Speculate on the future sustainable leadership ability of your subject.

6. Consider a corporation in the news and evaluate it in the light of the Jim Collins's (2001) dimensions of organizational greatness. Write an essay and determine those elements that appear consonant with the dimensions of Kouzes and Posner and those that do not. Speculate on the future sustainable leadership ability of your subject.

7. Go to the current business press and find an article on a company touted as an up-and-coming star. Analyze the company in terms of the good-to-great criteria discussed in the chapter and predict whether you believe the company's rise is sustainable or unsustainable. Write a short essay explaining your answer.

8. Go to the current business press and find an article on a leader touted as a up-and-coming star. Analyze the company in terms of the Kouzes and Posner criteria discussed in the chapter and predict whether you believe the leader's rise is sustainable or unsustainable. Write a short essay explaining your answer.

9. Consider the four gods of management. Which god best fits your personality. Write a short essay explaining why you believe this is so. Assuming your analysis is correct, write a paragraph explaining the implications of your observations for your career choice.

10. Consider *Good to Great* (Collins, 2001) and *The Leadership Challenge* (Kouzes & Posner, 2003). Write an essay highlighting similarities and differences between the principles mined by these empirical studies. Are the differences significant, and do the differences suggest the need for additional research or can the differences be reconciled by a change in perspective or emphasis?

Chapter 11

Assessing Technology Opportunities

11.1 ENTREPRENEURIAL ENGINEERS SEEK OPPORTUNITY

During the Cold War, there was a stark division of labor between business managers and the engineers who worked for them. Managers drove the business directions of the company, and engineers executed the design, manufacturing, and operational tasks managers assigned to them. Following the Japanese-inspired quality revolution, engineers have become more involved in pursuing and developing *opportunities* within and outside the organization.

To better understand the nature of opportunities and how to plan whether to exploit them or not, today's entrepreneurial engineer needs to understand, assess, and pursue opportunities, not only in terms of their technological feasibility, but also with a keen eye toward their business impact.

To do this, we examine the preparation of a *technology opportunity assessment* (TOA). TOA can be thought of as a miniature business plan (FastTrac, 2000). When an opportunity is potentially marketable (a marketable opportunity), a TOA is a preliminary step to preparing a more complete business plan. When an opportunity is viewed as largely helpful within an existing organization (an internal opportunity), a TOA is a preliminary step prior to investing significant funds in the idea.

To prepare a technology opportunity assessment, we need to consider:

- The opportunity
- Its competitive advantage, sustainability, and niche
- The economic case
- Preparation of the written TOA

The steps vary somewhat depending upon whether the opportunity is marketable or internal, but both types of opportunity should be subject to TOA scrutiny.

The Entrepreneurial Engineer, by David E. Goldberg
Copyright © 2006 John Wiley & Sons, Inc.

11.2   WHAT IS AN OPPORTUNITY?

Simply stated an opportunity may be defined as follows:

> *An opportunity is a correctable difference between the way things should be and the way things are.*

Parsing the words carefully, we note that opportunities are fundamentally about *change,* in particular, possible changes that should make life, in some sense, *better* for one or more individuals. Moreover, opportunities are about changes that can be *realized.* On the one hand, recognizing opportunity is an act of imagination, but it is not an act of utopian dreaming. For something to be considered an opportunity, it must have a realistic chance of achieving the desired change in the state of the world.

There are a variety of ways of classifying opportunities. Opportunities may be classified depending upon their sphere of influence and modality of effect. Commercial or entrepreneurial opportunities may be classified depending upon their locus within the enterprise's value chain. Each of these is briefly discussed.

Sphere of Influence

Opportunities may be classified based on their primary sphere of influence:

1. Personal
2. Entrepreneurial
3. Commercial
4. Societal

Each of these is briefly discussed in what follows.

Personal opportunities affect our personal lives. A potential job or career change, meeting a new person and possibly making a new friend, or considering a move to a new city are examples of personal opportunities. Although the methods discussed here can certainly be used in evaluating personal opportunities, they are not our primary concern and are not discussed further.

Entrepreneurial opportunities are those that involve the creation of a new product or service. Sometimes the new product or service may be provided by an existing enterprise, but oftentimes, a new product or service is launched through the creation of a new business entity.

Commercial opportunities occur in the course of running an extant enterprise; they involve a change to an existing product or service or the organization itself in a way that may improve the enterprise's competitive position, profits, internal efficiency, or other measures of commercial success.

Societal opportunities involve possible changes in public policy or action by private organizations in an attempt to improve particular conditions in society at large. The creation of a new charitable organization or the invention or modification of governmental social programs are designed to take advantage of societal opportunities. One of the pitfalls of societal or governmental opportunities is that it is difficult to know whether the imagined change can be realized by the proposed program or not; in business, at the very least, the sustained purchase of a profitable product or service by some set of consumers is a signal that someone, somewhere values the offering. Despite these difficulties of evaluating societal opportunity, the methods discussed here may be used in assessing such opportunities, but they are beyond the scope of our specific discussion.

Hereafter, we limit our discussion to commercial and entrepreneurial opportunities, their assessment, and how to proceed on the most promising of them.

Modality of Effect   Opportunities differ in their primary mode or means of implementation:

- Technological invention
- Organizational modification
- Financial engineering
- Creative marketing

Some opportunities are enabled through changes in technology or the discovery of new science. Others come about through organizational modification. Increasingly, opportunities arise through more efficient deployment of capital through the mathematics and methods of finance. Still others come about through improved communications or marketing. Many of the opportunities of concern to entrepreneurial engineers are primarily technological or scientific in nature, but oftentimes, opportunities in the real world involve a combination of these factors. We shall focus on opportunities that have a significant technological component.

Locus of Opportunity in the Value Chain   Commercial and entrepreneurial opportunities can be delineated by the primary locus of opportunity in the *value chain*. The term value chain has been used in the business strategy literature to distinguish common elements in a company's process flow. For example, Porter (1985) discusses primary value chain activities as follows:

1. Inbound logistics
2. Operations
3. Outbound logistics
4. Marketing and sales
5. Services

According to Porter, secondary activities include infrastructure, procurement, human resources, and technological development. Of course, entrepreneurial engineers may view technological development as more critical to the creation of a technological product or service than Porter, and, to be fair, Porter's analysis refers largely to going concerns with existing products and services. Nonetheless, it is useful to think of the enterprise systemically when seeking or trying to analyze different opportunities.

11.3 SUSTAINABLE COMPETITIVE ADVANTAGE: THE MAKING OF A GOOD OPPORTUNITY

In a competitive marketplace, enterprising individuals and companies are always on the lookout for new opportunities, and sometimes a new or existing company can get a head start on the competition. If the venturesome company is successful and the opportunity pans out, others will be attracted to it, however, and the advantage conferred by the head start can be quickly eroded by new competition or through the actions of important suppliers or customers. Knowing this, the entrepreneurial engineer will seek those opportunities that offer a *sustainable competitive advantage,* through the existence of one or more features of the product, service, or enterprise that are (a) different from those of competitive products and service and desired by customers, (b) difficult for the competition to emulate, and (c) difficult for suppliers or buyers to thwart. Here we divide our discussion into the four P's of competitive advantage and the five forces of sustainability.

11.3.1 Four P's of Competitive Advantage

The heart of competitive advantage is to *differentiate* a business from its competitors in a manner that is desirable or perceived to be valuable to its customers. To do this, introductory marketing texts talk about the big four control variables of marketing or the four P's:

- Product
- Place
- Price
- Promotion

Each of these is briefly reviewed.

In technology opportunity assessment, product is the king P of the four P's. Technology is used as a weapon to differentiate your product from other offerings, and it can be critical to have distinguishing features that customers will pay for.

The term *place* refers to where the business will be located, or more generally, what channels will be employed to get the product or service to customers. Companies selling software to the financial sector will do well to locate in New York, London, and other major financial capitals to be close to companies in their market. Companies that depend on recruiting specialized scientific and engineering talent will do well to locate near universities with excellent reputations in the needed talents. Of course, the virtual world of the Web has reduced somewhat the competitive advantage of physical presence or place. The Web permits customers almost anywhere to communicate, buy, and be serviced after the sale by companies that use e-mail, Web browsers, and other tools of online communication.

Price can distinguish a product from its competition, but startups should not usually aim at being the low-cost provider. Refining enterprise processes to make complex technological goods more cheaply than the big boys is capital and time intensive; neither of these variables favors a small business. If anything, startups and small businesses should aim to get a premium price for products or services perceived as "well worth it." High margins can cover a host of startup sins, whereas low margins can slow growth, delay breakeven, or worse.

Promotion or how a company penetrates a market with advertising, a sales force, manufacturers' representatives, or other means can be another way of creating competitive advantage. Engineers often fall victim to the *better mousetrap* theory of invention that says that customers will beat a path to your door if you simply invent a better gizmo. Engineers are surprised when they don't (and they usually don't), and selling your product or service must be taken seriously. Of course, no one is suggesting that heavy promotion will ensure a lousy product's success; however, a creative approach to promotion and market penetration can amplify important product differences and improve the perceived need for a product that might not otherwise exist.

Exploration Exercise

Consider a publicly traded company with products you are familiar with. In a short essay analyze the ways in which that company's combination of product–place–price–promotion gives it a competitive advantage over its competitors. Of the elements you discuss, identify which factor you believe to be the most important and why.

11.3.2 Five Forces of Sustainability

At a systems level, Porter discusses the *five forces* that make an industry or business opportunity more or less *sustainable* (Porter, 1980).

1. Rival of industry competitors
2. Threat of new entrants
3. Threat of substitute products or services
4. Bargaining power of suppliers
5. Bargaining power of buyers

A sustainable opportunity is one where each of these five forces or threats is diminished by the very nature of the industry.

The existence of a large number of intense competitors going toe to toe in an opportunity area is not good news. Even if those competitors have not recognized the particular opportunity, as soon as they get wind of competitive activity, they will seek ways to get into the same or similar business. If they are savvy competitors, they may already be working on something along the same lines.

If it is easy to enter a business, this does not bode well for sustained profitability. Even if there are no competitors now, ease of entry will increase the probability that others will match or exceed a product or service quickly. For startups, it is commonly important to have some intellectual property (IP) that protects the ideas at the core of the business. The IP can take the form of patents, copyrights, trademarks, or trade secrets, but having some core IP and protection for that IP is crucial. If a company relies on trade secret protection, the product or service (a) must not be easily reverse engineered and (b) employees, suppliers, and others with access to trade secret information must sign and be held to nondisclosure agreements. Larger companies often rely on government regulation, their size, large capital requirements, and economies of scale to discourage new entrants, but these barriers to entry are not usually available to the startup or small-business entrepreneur.

Products and services with many possible substitutes are not a good bet. Although substitutes may not directly compete in the same marketplace, their existence may severely limit what a company can charge for a product, thereby reducing profit margins, lengthening payback, and otherwise making the investment less attractive.

Businesses where suppliers or buyers have a lot of clout should be avoided. If, for example, a company depends on a single customer for most of its sales, that buyer may suddenly threaten to drop the product unless the company cuts its price substantially. The best businesses are those with large numbers of uncoordinated buyers and suppliers, leaving a company in a position to play suppliers off each other and receive good prices from its customers.

The foregoing discussion is most useful for those technology opportunities that involve starting a new product line or a new company; however, understanding how different internal opportunities affect a company's competitive position can also be helpful in selecting among competing projects vying for limited internal investment.

Exploration Exercise

Consider the industry of the publicly traded company you selected in the previous exploration exercise. Analyze whether the industry is favorable with respect to sustainability using Porter's five forces, and present your findings in a short essay.

11.4 WHAT IS YOUR NICHE?

The term *niche* is used in ecology and informally it denotes a species' "place" in the world. More formally, an ecological niche is sometimes defined as a region of a multidimensional space of environmental factors that affect the well-being of an organism. In commerce, businesses have niches, too, and by analogy a company's niche may be defined as a region in a space of competitive or environmental factors that affect the sustainable profitability of the business. In biology, physical location or geography plays an important role in determining an organism's niche, and likewise in business, where a business is located and where it distributes its products or services may be part of the niche, but as in biology, many other factors may go into the determination of a company's niche.

In biology, a principle called the niche exclusion principle is of particular interest here. The principle may be stated as follows:

> **Niche exclusion principle**   *In a competition between species that seek to exploit the same niche, only one species survives.*

This is quite stark, and when transferred to the business context, it highlights the importance of finding a niche. If a company goes after the same set of resources in the same neighborhood as a stronger competitor, given enough time, the company is likely to lose out.

Companies, like species, like to have large, resource-full niches all to themselves. By paying attention to the five factors and by considering the key elements to their profitability, companies carve out high profit margins and keep the amount of head-to-head coverage down to a manageable amount. Even companies in the same business do this. For example, in the automotive business, after the onslaught of Japanese competition in the family compact cars and sedan market, American manufacturers took refuge in the production of minivans, SUVs, and light trucks. Of course, American manufacturers did not give up on compact cars and sedans, but when large, powerful, longstanding companies struggle to carve out and profit from specialized niches in a market, the startup or smaller firm, with fewer resources and almost no economies of scale, must be clear about its niche and be convinced of its sustainability for some reasonable period of time.

A useful tool to help analyze an opportunity's niche is the so-called *positioning diagram*. Positioning diagrams (sometimes called *perceptual maps*) plot the imagined or measured location of a product or service on two or more attribute axes. For example, we might imagine the battle between automakers being played on competitive dimensions of price and function as shown in Figure 11.1. At the top of the price range are luxury vehicles and at the low-end are affordable cars. On the left end of the lifestyle dimension are conservative transportation automobiles, and on the right end are sporty vehicles. When consumers are asked to rate their perceptions of different automobiles, they locate them in different parts of the price–lifestyle space. High-performance German brands such as BMW and Porsche tend to be placed high on the price (luxury) scale, and they tend to be rated as high performance. American sedans such as Dodge and Ford, tend to be rated as lower priced and more conservative.

In trying to understand the competitive position of a new product or service, it is often difficult to obtain real data, but if the product or service is going into an existing market, positioning studies may be available with data regarding existing competition. Moreover, using such drawings as conceptual tools to understand whether an opportunity has a defensible niche is a common and useful exercise.

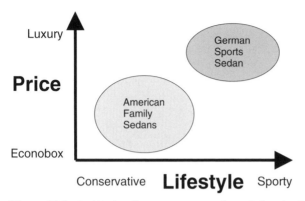

Figure 11.1 Positioning diagram or perceptual map helps visualize a product's niche.

Exploration Exercise

Consider a publicly traded company with a product you are familiar with. Choose two key factors that effect consumer choice with respect to this product. Sketch a positioning diagram, locating the selected company's product with respect to its competitors.

11.5 THREE FINANCIAL MYSTERIES OF OPPORTUNITY ASSESSMENT

In earlier times, when engineers were employed strictly on technical matters, their lack of understanding of key tools of financial communication and analysis was not important. With the rise of the entrepreneurial engineer, it is increasingly important for engineers to read financial statements, think relentlessly in terms of costs, prices, value, and breakeven, and to better understand the time value of money. Here we seek a short introduction to these items, although deeper coverage should be sought in an up-to-date text on engineering economics or managerial accounting (Park, 2002).

11.5.1 Overcoming the Fear of Financials

Engineers who have not seen them before can find financial statements quite daunting. At first, this "fear of financials" seemed odd because the mathematics employed in preparing financial statements rarely goes beyond basic arithmetic. However, over time, I've come to realize that the fear is born of a lack of understanding of key financial terms and principles underlying the statements. Although a complete introduction to financial statements is beyond the scope of our treatment, we concentrate on the following four items:

1. Cash flow statements
2. Balance statements

3. Profit-and-loss (income) statements
4. Ratio analysis

Each is briefly discussed below, but readers desiring a compact, yet thorough introduction to the topics should refer to Tracy's useful book (Tracy, 1999).

Cash Flow Is King

Ultimately, a new company wants to make profits, but cash is the mother's milk of starting a new business and is initially more important than profits. Even after a company becomes more established, if its cash position is neglected, it is difficult to meet payrolls, pay suppliers, and otherwise keep the business afloat despite theoretical calculations predicting a year-end profit.

Cash flow statements are the easiest of the three main financial statements to understand because they are *just like a checkbook*. The amount of money coming in, less the amount of money going out, equals the net increase or decrease in cash from all sources.

In reading financial statements, there is a convention that is almost universally used. Conventionally, engineers express positive quantities with a plus (+) sign or no sign at all and negative quantities with a minus (−) sign. Accountants express positive quantities with no sign at all and negative quantities inside of a pair of parentheses. Thus, using the accounting convention, $100,000.00 and ($100,000.00) are positive and negative one hundred thousand dollars, respectively.

Profit and Loss: Income Statements

Checkbook thinking is fine for running a household or for running a small proprietorship, but calculating profits for a large business on a checkbook or a *cash basis* can be severely misleading because, by their nature, incoming cash flows tend to be slow in arriving and outgoing cash flows are delayed as long as possible past the time of purchase. Moreover, large expensive investments in machinery and other capital equipment have a long useful life, and it is desirable to charge those expenses over multiple years. Therefore, most businesses larger than a mom-and-pop store calculate their profits on an *accrual basis.* That is, sales and normal expenses are recorded as soon as they are booked regardless when they are paid. Capital investments are *depreciated* over time, and a depreciation charge is taken each year to spread the cost of the expensive equipment over its useful life.

Keeping Your Financial Balance: The Balance Statement

Cash flow and income statements identify the *flow* of funds through a business in somewhat different ways. The balance statement is a *snapshot* of the *distribution* of assets, liabilities, and ownership (net worth or equity) at a given point in time.

These three things are related through the *accounting equation*:

$$\text{Assets} = \text{Liabilities} + \text{Owners' equity}$$

Assets are items of value possessed by the company or commitments by others to pay the company some amount. This latter quantity is called an *account receivable* because it is an amount that the company expects to receive in the future under current obligations.

Liabilities are current or long-term commitments to pay off loans, notes, or bonds in the future or current promises to pay others for goods or services ordered on account. This latter type of quantity is called an *account payable* because it is an amount that the company expects to pay as bills come due in the future.

The difference between assets and liabilities is then the owners' equity or the financial net worth of the company.

Ratio Analysis: Dimensional Analysis with Money

Engineers commonly use *dimensionless ratios* such as the Reynolds number and the lift coefficient to help understand complex experimental data, organize equations and model building, and relate model testing in a wind tunnel to that of a prototype flight. This kind of analysis is called *dimensional analysis* because it uses the dimensional nature of important variables to derive dimensionless parameters that determine the scaling properties of a physical system.

In financial matters, a similar technique, *ratio analysis*, is used to judge whether a company is healthy with respect to industry standards. In ratio analysis most of the quantities compared are quantities of money, so the ratio of one quantity of money to another quantity of money is, by definition, a new dimensionless ratio (like the Reynolds number or lift coefficient), and these financial ratios can be used to compare businesses in similar industries or businesses (in engineering, such comparison is referred to as *dynamic similitude*).

Perhaps the most familiar ratio to the ordinary investor is the so-called price-to-earnings (P/E) ratio:

$$\text{P/E ratio} = \frac{\text{Current market price}}{\text{Earnings per share}}$$

P/E values are used to judge whether a stock is a good value relative to its earnings stream. Another way to think of the P/E is to take its reciprocal and think of that number as an interest rate. This interest rate interpretation would be completely valid if the entire amount of the company earnings per share were paid out as a cash dividend. For example, a P/E of 20 can be thought of as comparable to an interest rate of $1/20 = 5\%$.

Various ratios measure profits relative to investment and sales. Two important ones are *return on sales* (ROS) and *return on equity* (ROE):

$$\text{ROS} = \frac{\text{Net income}}{\text{Sales revenue}}$$

$$ROE = \frac{\text{Net income}}{\text{Shareholder's equity}}$$

Other ratios measure creditworthiness. One such measure is the debt-to-equity ratio:

$$\text{Debt-to-equity ratio} = \frac{\text{Total liabilities}}{\text{Total stockholder's equity}}$$

Generally, lenders like to loan money to those who have more to lose in bankruptcy than they do, and debt-to-equity ratios less than 1 ensure this.

Exploration Exercise

Consider a publicly traded company with products you are familiar with. Obtain a copy of the company's annual report. Examine and interpret the cash flow, income, and balance statements. Is the company profitable? Did the company's cash position improve or degrade last year? Did stockholder equity improve or degrade last year? Calculate the ratios discussed in this section. Using library data or data off the Web, compare the companies averages to industry averages. In what ratios is the company's position favorable or not.

11.5.2 Prices, Margins, and Breaking Even

Financial analysis of new technology can get fairly sophisticated, and detailed business plans can contain a variety of means of projecting revenues, but the most useful starting point for thinking about the finances of a technology opportunity is simple breakeven analysis.

Simple breakeven analysis turns on building a simple linear model relating various costs and price. Say we get a price P per unit of product for which we pay a fixed cost F and a variable cost V. If we sell n units of the product, we may calculate the net revenue R received as follows:

$$R = nM - F \tag{11.1}$$

where the contribution margin is defined as the difference between the unit price and the variable cost $M = P - V$.

Breakeven occurs at that volume n_b where net revenue $R = 0$. This results in a relationship between breakeven volume, fixed cost, and contribution margin as follows:

$$F = Mn_b \tag{11.2}$$

or solving for n_b,

$$n_b = F/M \tag{11.3}$$

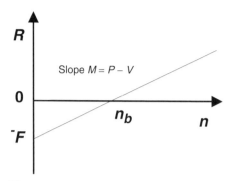

Figure 11.2   Breakeven analysis is illustrated by a plot of revenue versus volume. At zero volume, the fixed costs are fully charged. At breakeven, net revenue is zero, and above breakeven, the margin accumulates straight to the bottom line.

Finally, total revenue may be written in terms of M and the breakeven point by substituting Equation (11.2) into (11.1):

$$R = (n - n_b)M \tag{11.4}$$

In words, the net revenue can be calculated as the excess volume over breakeven times the unit contribution margin.

In all these calculations, the key quantity is, not so much the price or the variable cost itself, but rather their difference, the contribution margins. All efforts to widen margins result in a concomitant reduction in breakeven point and increase in net revenue.

Graphically, the situation is illustrated in Figure 11.2, where net revenue is plotted as a function of volume. Of course, the mathematics here is nothing fancy, but the beauty of breakeven analysis is in the distinction it makes between fixed costs and variable ones. By making that distinction, you can calculate roughly and quickly to better assess opportunities as they arise.

11.5.3   Time Value of Money

Simple breakeven analysis is quite useful for getting a ballpark sense of when a given investment will start to payoff, but more sophisticated analysis will take account of the alternative uses of money and the fact that a dollar today is worth more than a dollar tomorrow.

The key idea is to take a present amount P and account for the interest i that will compound over n periods to give some future amount F as follows:

$$F = P(1 + i)^n \tag{11.5}$$

The validity of the equation is easy to verify as in one period the amount P will be increased by the multiple $1 + i$. In the next period it is increased by an identical factor and so on for n periods.

Almost all sophisticated financial analyses of opportunities or various deals start from this point of departure. Getting used to "moving" sums of money around in time through the use of Equation (11.5) will help you better understand this kind of analysis.

11.6 WRITING THE TECHNOLOGY OPPORTUNITY ASSESSMENT

In assessing technology opportunities, it is important to understand what the opportunity is, whether it represents a sustainable competitive advantage, the market for the opportunity, a preliminary financial analysis, and a discussion of personnel who are key to exploiting the opportunity.

The elements of a TOA are fivefold:

1. Executive summary
2. Opportunity description
3. Preliminary market analysis
4. Preliminary financial analysis
5. Action plan

Each of these can be further discussed.

11.6.1 Executive Summary

The executive summary is a one- or two-page summary of the main points in the body of the TOA. A good format is to use *titled paragraphs or sections,* where a sequence of paragraphs or sections are headed by a headline in a bold font.

Good executive summaries are hard to write because they need to be brief yet contain the key ideas in a form that is easily and quickly grasped. A good process to follow is to write the individual sections of the opportunity assessment first, and then work on condensing the larger sections down to their essentials.

A key to this condensation step is to think primarily in terms of what the decision maker or investor needs to know to make a decision in your favor. Thinking in these terms usually (a) eliminates technical detail unrelated to competitive advantage, (b) boils financials down to aggregate numbers and key profitably statistics such as time to breakeven and return on investment, and (c) increases the use of familiar products and companies as analogies.

This condensation step should be taken a step further in the preparation of an *elevator pitch.* Imagine you are riding in an elevator with a famous venture capitalist (VC). You have one minute to interest this person in your idea before the elevator door opens. What would you say? Probably you need to tell the VC the following:

- Company/product name and location
- The fundamental market problem the company solves

- How the company/product overcomes the problem
- The sustainable competitive advantage the company has over other approaches

In other words, the elevator pitch takes your executive summary and condenses it down to a single short paragraph. If you thought the condensation from full business plan or opportunity assessment or executive summary was hard, the condensation down to an effective elevator pitch can be excruciating, but it is necessary. Busy people don't have long periods of time to assess whether what you are doing is interesting or not. They decide quickly, and their default decision is "no." Learning the art of the executive summary and elevator pitch are important to getting your ideas accepted by busy decision makers and investors.

Exploration Exercise

Consider a technology opportunity that might lead to a new business. Do a preliminary Web search to see if there are companies doing something similar. Write an elevator pitch for the idea as a prelude to preparing a full TOA.

11.6.2 Technology Description

A key element of any technology opportunity assessment is the description of the technology. In doing so you need to cover the following elements:

1. The type of opportunity
2. The purpose and function of the technology
3. The current stage of development of the technology and additional work required to commercialization
4. Barriers to imitation
5. Regulatory and legal concerns
6. Implementation or production of technology

Each of these is briefly considered.

Type of Opportunity By "type" of opportunity, the reader of the TOA wants to know whether the opportunity is

- Internal
- External
- Startup

An internal opportunity is one where internal processes or procedures in an existing organization are modified to realize some reduced cost, improved quality, or some other benefit. An external opportunity is where an existing corporation modifies an existing product or service or starts a new product or service. A startup opportunity is where a new business entity is created to launch a new product or service.

Purpose and Function of the Technology   In this section, describe what the technology is supposed to do for the user or customer and give a concise yet qualitatively complete description of how the technology works.

Current Stage of Development and Work Required to Commercialization   Is the idea merely a concept, or has pilot or prototype development occurred, and how far along is that development? What further effort is required to create a version that may be used or sold?

Barriers to Imitation   How is the idea protected from imitation by competitors? Is it protected by intellectual property law through patent, copyright, or trademark? If not, is it difficult to imitate by its nature? How so? Is it difficult to reverse engineer? Is it protected as a trade secret? Are there unique, exclusive sources of components or other supplies? Are unique labor inputs required to develop, maintain, support, and extend the product or service?

Regulatory and Legal Concerns   Does the technology fall under a governmental regulatory regime? In what ways? What government licenses or permission must be obtained before the product can be manufactured and sold? What potential uses or misuses of the product or service will open the company to product liability or other lawsuits?

Implementation or Production of the Technology   How will the technology be implemented or manufactured? Will an in-house capability be developed or will manufacture be outsourced?

 Exploration Exercise

Consider a technology opportunity that might lead to a new business. Write an opportunity description for the idea.

11.6.3 Market Analysis

Describing your gizmo is the easiest part of the technology opportunity assessment exercise, but now the going gets tougher. A critical part of the analysis is to assess whether there is a market for what you want to do. For an internal opportunity, your "market" is some "customer" within your own organization, and a formal market analysis may be unnecessary. For a commercial or startup opportunity, however, it is essential to do a good job in collecting facts and figures about your market, and then presenting that data to show there is a sufficient market to exploit your opportunity.

Incomplete List of Sources for Market Data

There are many sources of market analysis data:

- Trade magazines, associations, and websites
- Publicly traded company data and corporate websites
- Government and private statistical sources
- Privately prepared market research reports

Established industries almost always are served by a trade press and many have formal trade associations, some with websites and magazines. Consumer markets are often driven by demographics and *American Demographics* (www.demograph ics.com) is a useful reference. If your competitors are publicly traded, their Securities and Exchange Commission (SEC) filings have useful data (www.10kwizard.com), and many companies have useful information on their websites. The U.S. government has a wealth of statistical data, and most of the useful sites can be reached from the website www.fedstats.gov. Dun and Bradstreet (www.dnb.com) and Hoovers (www.hoovers.com) are useful sources of corporate information, although useful data comes at a price. There are a number of firms that provide market research reports on different segments of various markets for a price and Web searches should reveal some of the best known quite quickly. Business media outlets (www.bloomberg.com and www.wsj.com) are useful sources of information as well.

Writing the Market Analysis

Collecting market data for the first time can be an enlightening experience for entrepreneurial engineers, but organizing that data into a coherent market analysis is a challenging task. To make that task easier, we outline a number of elements of the analysis in a sequence that will lead to a sophisticated analysis. In particular, we suggest working from the bottom up, starting from the product to the individual customer to the market segments that might benefit from your product:

- Analysis of features, benefits, and consequences (FBC)
- Analysis of customers
- Analysis of competition
- Most promising market segments, size, and gradient
- Market penetration and promotion

Each of these is briefly discussed.

FBC Analysis

Although the first section of the TOA analyzes the opportunity in technological terms, the market analysis must revisit the product or service through the *eyes of the customer*. In particular, the analysis of features, benefits, and consequences considers what the product or service does and how that benefits the customer directly and in the long term. An example from a common software product will make this clear.

Consider a word processing program. A common useful *feature* is the built-in spelling checker. How do spell-checks *benefit* a customer? They reduce the number of misspelled words in a document, and they reduce the need to consult a dictionary as often when you are uncertain about the spelling of a word. Longer term, what are the consequences of these benefits? Documents with good spelling may improve business for the user because bad spelling is viewed as a sign of carelessness. The time savings of not having to consult a dictionary as frequently will make a user more productive, and the time savings can be used to spend more time with loved ones, a hobby, or to get more work done.

By thinking deeply about the features of a product as well as their benefits and consequences for customers, you build the foundation for your entire marketing plan.

Customer Analysis

Having thought about how the product affects customers through FBC analysis, it is time to consider the customers themselves. In particular, we decompose the customer analysis along three dimensions:

1. Demographics
2. Role analysis
3. Decision paths

Each of these is considered briefly.

Demographic analysis considers the age, cultural, and gender distribution of the customer. In consumer products, demographic analysis is crucially important, but even in more technical goods and services, knowing the demographic distribution of your target market can be helpful in product design and the design of market penetration campaigns.

Role analysis examines the type of role being played by your customers, and we can distinguish between *titular* and *purposive* roles. Titular roles are those associated with the usual *titles* people possess in the job market. For example, is the customer a CEO, an engineer, a clerk, or a factory worker? Knowing your customer along these lines tells you a lot about their educational background, level of technical expertise, and what's important to them in making a buying decision.

Analyzing a customer's purposive role tells us the broader purpose behind his or her buying decision, and we can distinguish between three broad categories: utility, instrumental, and display. Ordinary consumers of goods and services seek utility directly. For example, a buyer of a Krispy Kreme doughnut buys a fat-filled confection and benefits (so he or she believes) from the sensory stimulation (and immediate weight gain) of consumption.

For many technical goods and services, the customer buys the product as an *instrument of success* in some other enterprise. If you make components that others can use, customers buy those components to add value to their larger system. Finally display customers buy things, at least in part, to show them off to others. Luxury consumer goods fall into this category, and the status conferred by displaying symbolic trademarks can be as important as the functionality of the product. Technical goods and services usually require utility or instrumentality, but it can be beneficial to pricing and margins if technical goods and services customers are motivated at least partially on display grounds.

A crucial element to marketing is understanding the various paths of buying decisions. Complex technical goods often are *recommended* to upper management by low-level engineers or technicians. Depending on the organization and the cost of the goods, the actual decision might require a whole host of signatures up the chain of command. In selling goods and services, it is important to have access to the ultimate decision maker, but it is equally important to understand the chain of recommenders that leads to that person's desk.

Competition Analysis

Creating and maintaining a sustainable competitive advantage requires that you fully understand the direct and indirect competition for your technology opportunity. What firms and products and services compete with your technology? What market segments do these firms supply? What share of those market segments do these companies currently occupy? How do competitor products and services stack up on FBC terms with your proposed offering. How do your customers price their product? Along what dimensions is your offering unique and can that uniqueness be represented with a positioning diagram or perception map?

Most Promising Markets and Segments

Given your product/service offering, your customer, and your services, what are the most promising markets or market segments for you to target? How large are they and how fast are they growing?

Market Penetration and Promotion Analysis

Given you offering, your customers, and markets, what are the most effective means of reaching your target audience?

- Dedicated sales force
- Manufacturers representatives
- Direct mail
- E-commerce

- Telemarketing
- Piggyback sales (bundling your product into another)

These categories discuss the means by which you *actively* mean to close sales, take orders, and deliver product.

Additionally, there a number of ways that you intend to *inform* potential customers of your offerings as preparation to closing sales. Specifically, what means of advertising and publicity will reach your target customers most effectively?

- Trade advertising
- Mass media advertising
- Trade shows
- Catalogs
- Web-based promotion
- Press releases

In a short time, Web-based approaches have become a great equalizer in marketing and promotion. Although other traditional methods of reaching customers must be exploited, it is the rare new business today that does not use the Web to some degree in penetration and promotion efforts.

Exploration Exercise

Consider a technology opportunity that might lead to a new business. Investigate and write a preliminary marketing analysis for the idea.

11.6.4 Preliminary Financial Analysis

In writing a preliminary financial analysis, the main point is to determine whether the business has sufficient margins and volume to justify a business. This can range from a simple back-of-the-envelope breakeven analysis to a more complete set of 3-year pro forma. More formal texts on entrepreneurship (Timmons, 1999), business planning short courses and workbooks (www.fasttrac.org), and business planning software (www.bplan.com) can be used on the more formal end of this planning spectrum.

At a minimum, you need to estimate the financial picture surrounding your budding opportunity. A 3-year time frame is a good starting point, and income and outflow should be estimated in the following categories:

- Product sales volume and pricing
- Research and development expenses
- Marketing, promotional, and sales expenses
- Administrative expenses

The art of estimating volume and pricing for a new product is quite difficult. Knowledge of competing or similar products can be quite helpful in projecting reasonable prices, initial volumes, and reasonable growth rates. Most investors will not take wild guesses at face value, so use of historical data in an existing market or analogous market is almost essential.

Research and development expenses include R&D (research and development) salaries, licensing fees (for technology owned by others), patent filings, equipment leasing, and other expenses surrounding the protection of intellectual property. Salaries, legal fees, and other expenses can be estimated fairly accurately; the difficulty in giving accurate estimates of R&D expense estimates are usually tied to the uncertainty inherent in estimating the amount of time and effort required to get to a marketable product. A technology coming from a university or other laboratory often requires a fair amount of development to make it suitable for use in a product. The engineer's natural tendency to tinker and to raise the specifications bar as more is learned exacerbate this problem.

Exploration Exercise

Consider a technology opportunity that might lead to a new business. Investigate and write a preliminary financial analysis for the idea.

11.6.5 Action Plan

To this point, the technology opportunity assessment has been an exercise in exactly that, assessment. The final section of the TOA discusses the most promising options going forward, and it should contain the following information:

- Discussion of most promising options
- Projection of financing required and potential sources
- Analysis of key roles and personnel

Each of these is briefly considered in what follows.

Selection of Most Promising Options

The purpose of the TOA is to help you to decide what to do with the technology opportunity at hand. Here you make a list of one or several options that appear to be most promising and evaluate your current assessment of likely success for those options. Typical options include the following:

- Move ahead with technology opportunity.
- Write a full business plan.
- Seek licensing opportunities for the technology.

- Seek additional input on the feasibility of the idea.
- Continue development of technology outside commercial arena.
- Shelve the technology or discontinue continued evaluation.

Each of these is discussed.

The most aggressive move is to move ahead with the technology by starting a new company or starting a new business within an existing company. Normally, TOA information is insufficiently detailed for this purpose; however, a particularly well-researched TOA or a particularly hot idea may warrant moving ahead without the formality of a full business plan.

The usual way to proceed with a promising TOA is to write a full business plan as a preparatory step to doing the startup. Normally, TOAs do not contain sufficient market, financial, executive team, corporate structure, or corporate operational detail to launch a new business, and writing a formal business plan starts from the TOA and adds the necessary detail.

The TOA process may reveal commercial possibilities that the assessment team is not interested in pursuing themselves. In cases such as this, it may make sense to explore the possibility of licensing the technology to others. The TOA has probably identified key players in existing and related markets, and that list may be a good list of licensing prospects.

Despite thorough research, technology opportunity assessment may come back with an inconclusive assessment. In those cases, the technology may not quite be ripe or the markets may not yet be sufficiently developed. In cases such as this, it may be useful to shop the TOA around to potential investors, academics, and potential customers, not with the idea of getting an investment, but with the objective of getting feedback on how to reshape the idea so that it is more obviously promising.

Occasionally, a TOA will reveal ways in which the technology must be improved before it is marketable. This suggests returning to the lab to improve the technology along those lines; however, engineers, even entrepreneurial ones, have a predisposition toward playing around with and fine tuning their gizmos. If your assessment suggests going back to the lab, make sure the reasons are sound. An imperfect technology with a strong competitive advantage today is far superior to a perfected technology entered into an already established marketplace.

Finally, one usually undertakes a TOA with the idea of moving ahead somehow, but data is data, and the assessment may reveal any number of deal breakers:

- Extant technology in the market with comparable features
- Insufficiently large market to build a profitable business
- A technology with competitive advantage that is not sustainable

Each of these argues against going ahead with the opportunity, but the determination of viability along these dimensions is subtle.

The existence of competitors does not automatically preclude moving ahead. Perhaps the competitors have been lazy or have not kept up with advances in technology. You seek competitive advantage, not necessarily monopoly.

Many venture capitalists like to see a billion-dollar market before backing a technology idea, and perhaps your TOA doesn't add up to something that sizable. Perhaps you have undervalued the potential for growth of the market, or perhaps you have ignored market segments that could add up to sufficient size.

Technology with a current competitive advantage that enters a ruthless market of rampant rivalry, ease of entry, power exerted by suppliers and buyers, and many substitutes is in some ways asking for future trouble. The technology may have a temporary competitive advantage initially, but market conditions may cause a variety of pressures that erode that advantage. Even in these daunting circumstances, going ahead may be possible if you can imagine a way to stay ahead of the pack. Technology, by its very nature, is a dated product. Anyone entering a technology-based business is either implicitly or explicitly signing up to live a life of continual improvement and innovation. Perhaps one approach to such a competitively relentless pursuit is to think like a chess player. Plan out your opening, but also think through your strategy and tactics for the middle game as well. If you can imagine a sequence of innovations that will keep you ahead of the competition, your chances for longer term survival will improve.

Analysis of Key Roles and Personnel

Technology inventors are a proud bunch, and it is often the case that the key technologists are important players in getting the technology opportunity off the ground. It is possible for inventors, however, to overestimate their importance in the larger scheme of things, and there are many other key players in a technology startup.

- Chief executive officer
- Chief financial officer
- Chief technical officer
- Sales manager

The technology inventor will, for example, often be tempted to take on duties as CEO along with technology development responsibility, but this can be a mistake unless the inventor has significant prior business experience. A CEO with a track record of entrepreneurial success and a current Rolodex full of friends in strategic places can organize routine business functions, access funding, make initial product placements, and do other important tasks more quickly and better than a novice company president doing expensive on-the-job training.

The chief financial officer (CFO) keeps tabs on where the money is coming from and going. Despite continuing financial shenanigans by CFOs at a number of high-profile companies, the traditional role of a CFO is to be a company's chief reality therapist. Good and bad news is best reflected in the financial statements, and a good CFO keeps the growth plans and expectations in line with market reality.

The chief technical officer (CTO) is responsible for technical research and development. This is a position that many inventors will assume in startups, but taking the title of CTO requires a level of managerial responsibility that the lone inventor is not ready for. Many technically based companies will assign the founding inventor a position as technological guru or thinker-in-chief that requires a maximum of inventiveness and a minimum of line managerial responsibility.

Sales are the lifeblood of all commercial enterprises, and the care and feeding of a sales force, and the selection of a sales manager, can be a mystery, pure and simple, to the average engineers. By the nature of the business, salespeople are extraverted, intuitive, and brash, and this can be a clash of cultures with introverted, rational, and conservative engineering personalities. Understanding these personality and cultural differences is important.

Exploration Exercise

Consider a technology opportunity that might lead to a new business. Write an action plan for the idea.

Projection of Financing Required and Identification of Potential Sources

Many of the most promising options will require significant levels of funding to proceed. The preliminary financial analysis will help estimate the cash needs to get started, but the entrepreneurial engineer's tendency toward optimism will, in this case, tend to underestimate cash needs.

Funding can be raised from a variety of sources (in rough order of increasing cost and increasing loss of control):

- Personal funds of the startup team
- Friends and relatives of the startup team
- Entrepreneurial incubators
- Angel investors
- Venture capitalists
- Initial public offerings
- Merger

The first four of these occur early in a company's development, and the latter three occur somewhat later.

The initial funding of a new technology business will often come from the inventors, their startup team, or personal friends and relatives. This is the iconic "garage" startup of Hewlett-Packard and Apple legend. More and more, entrepreneurs are being supported by incubators on university campuses or elsewhere with seed funding, low-cost office space, and experienced business advice. Beyond incubators is an important class of funding from so-called *angel* investors. Angels are wealthy, knowledgeable investors, often entrepreneurs themselves, who have money to invest, often in businesses related to their own personal technical or business expertise. Angels generally are willing to invest smaller amounts earlier than so-called venture capitalists (VCs).

Venture capital firms are professional organizations that raise funds to invest in startups well beyond the incubation stage. Beyond these private sources of funds, companies can raise money by selling stock to public stock markets through *initial public offerings* (IPOs). Public offerings are not usually possible until a company has a track record of successful products and profits. Finally, there can come a point where a company believes its financing and market access is fundamentally limited by its size or market conditions. In such cases, the financing for growth can come from a larger company that acquires your firm. Oftentimes, the larger company exchanges shares of its stock, cash, or some combination of cash and stock to acquire the smaller firm.

Exploration Exercise

Consider a technology opportunity that might lead to a new business. Assemble a full technology opportunity assessment from previously written pieces. Write an appropriate executive summary, and prepare a 10-minute presentation on the idea for presentation to an group of potential investors. Make the presentation.

SUMMARY

The chapter started by considering the changing role of engineers within companies and organizations today. Where once engineers were largely responsible for demonstrating technical feasibility, they now are more intimately involved in the search for and exploitation of opportunity. These new responsibilities require that engineers understand their organizations from a systems point of view. Specifically, this systems viewpoint requires a more sophisticated approach to customers, markets, finances, and their interrelationship to the technology itself.

The chapter defined opportunity as a "correctable difference between the way things should be and the way things are," and this definition highlights three things. Opportunities are about (1) change, for the (2) better, in a (3) realizable manner. The chapter categorized opportunities depending upon their sphere of influence, modality, and locus. Sphere of influence relates to *what* opportunities affect. Are they personal, entrepreneurial, commercial, or even societal? Modality refers to *how* an opportunity creates change. Is it done through technological, organizational, financial, or marketing means? Locus refers to *where* opportunity exists in a company's value chain. Is the opportunity an improvement in supply or delivery of product? Is it a manufacturing innovation, a marketing innovation, or a services innovation?

A key concept in evaluating opportunity is the notion of sustainable competitive advantage. The competitive advantage side of this comes by differentiating the product or service delivered along the four P lines: product, place, price, or promotion. The sustainability of the competitive advantage is often examined through the lens of Porter's five forces. Those opportunities with modest rivalry, low threat of new entrants, difficulty in substituting other products or services, and poor bargaining power of suppliers and customers are most likely to be sustainable in the long run. Opportunities threatened along one or more of these dimensions are depending upon successfully outflanking their competitors, suppliers, and customers over the long haul.

The ecological notion of a niche was defined and carried over to the business context. In ecology, niches are sets of environmental factors that affect the well-being of an organism, and the same definition carries over to business if we simply substitute the term *business* for the term *organism*. The niche exclusion principle in biology suggests that in the long run, there will only be one survivor in a particular niche between two or more organisms. In business, this places a premium on some kind of differentiation of your product from your competitors. The chapter discussed the idea of a positioning diagram or perceptual map to help visualize niches graphically.

The chapter considered the engineer's "fear of financials" and attempted to overcome it with a straightforward discussion of cash flow, income, and balance statements. The mathematics of these statements is simple arithmetic, but the confusion seems to come from not understanding a few key terms. The chapter also considered the utility of ratio analysis—what engineers might call dimensional analysis for companies—breakeven analysis, and the basics of the time value of money. Although a full discussion was beyond the scope of this chapter, the discussion served up the basics and may also serve to introduce more careful treatments in other courses or texts.

The chapter concluded by outlining the preparation of a written technology opportunity assessment, or TOA. TOAs should have a clear executive summary, a description of the technology opportunity, preliminary marketing and financial analyses, and an action plan or recommendation. TOAs are difficult documents to write because they range broadly across a system as an enterprise yet focused on making a business case for investors or other decision makers TOAs can be the basis of direct action, a more detailed business plan, or they can lead to more development or shelving of the technology. From the standpoint of the entrepreneurial engineer, trying your hand at writing a technology opportunity assessment can be one of the quickest ways to become more familiar with the complex interplay between technology and the enterprise it enables and serves.

EXERCISES

1. Read the *Wall Street Journal* for one week, collecting one article each day that discusses technical opportunities that various companies are considering or exploiting. Try to choose articles from the same or a related industry. In each of the articles identify key points of competitive advantage for the particular opportunity and in a short essay discuss whether those advantages are likely to be sustainable using Porter's five forces analysis.

2. Select an article in the business press that discusses a company that is in financial difficulty or bankruptcy. Identify (a) how the company got into financial difficulty, (b) what steps it is taking to get out of financial difficulty, and (c) whether additional funds are being injected into the firm and by whom.

3. Many methods of using the Web for marketing were tried in the early days and some of them have survived. Identify five websites that use different approaches to using the Web to reach customers. Write a short synopsis of each approach and assess whether the same or similar approach can be used for promoting a technology-based opportunity.

4. Over the course of 2 months, "adopt" a publicly traded company by following its stock price, annual report, financial news, products, SEC 10 K filings, industry analyst reports, and other data. Is the firm relatively strong with respect to its industry? Why or why not? Were the reasons for relative success or failure apparent at the point of decision? In hindsight were there any steps the company could have taken that would have improved its current situation?

5. Repeat Exercise 4 using one of the company's listed in Jim Collins's (2001) book *Good to Great*. Use one of the good-to-great companies or the reference not-so-great companies.

6. Investigate a university near you for campus resources designed to aid entrepreneurs. Write a list of resources, services, funding sources, and other aid that the university can help you obtain.

7. Read the Web pages of three venture capital firms. What businesses have they invested in and at what stage? Compare and contrast the firms based on the information provided on the website.

8. A properly investigated technology opportunity assessment is a substantial undertaking, requiring library and Web data collection, analysis, creative problem solving, and thoughtful projections. Prior to investing that kind of time, it is sometimes helpful to do a prescreening on the basis of more cursory investigation. A useful pre-TOA analysis focuses on (a) the idea, (b) competition, and (c) a back-of-the-envelope breakeven analysis.

9. A common technique for inventors and other creative problem solvers is to create a *bug list* of nagging problems or difficulties in the world that might be deserving of a clever and profitable solution.

10. Attend a trade show in an area of commercial interest to you. Observe the array of companies present, product and service offerings, promotional activities, and the interaction of trade show vendors and customers. Write a short memo listing and analyzing your observations.

11. Obtain a business plan on the Web at www.bplan.com or comparable site. Write a critique of the style and substance of the business plan.

12. Make a list of websites helpful to technology opportunity assessment. Start with the Kauffmann foundation's website www.entreworld.org.

13. Jim Collins, author of the bestseller *Good to Great,* has written that social or organizational innovation is often more important to business success than technical innovation. Write an essay debating the merits and shortcomings of this argument.

References

ASCH, S. E. (1951). Effects of group pressure upon the modification and distortion of judgement. In H. Guetzkow (ed.) *Groups, leadership and men*, Pittsburgh, PA: Carnegie Press.

AXELROD, R. M. (1984). *The evolution of cooperation*. New York: Basic Books.

AYER, A. J. (1946). *Language, truth, and logic* (2nd ed.). London: Victor Gollancz.

BACON, F. (1994). The great instauration. In E. A. Burtt (Ed.), *The English philosophers from Bacon to Mills* (pp. 5–24). New York: The Modern Library. (Original work published 1620)

CARNEGIE, D. (1981). *How to win friends and influence people* (rev. ed.). New York: Pocket Books.

CHANDLER, A. D. (1977). *The visible hand: The managerial revolution in American business*. Cambridge, MA: Belknap Press of Harvard University Press.

COASE, R. H. (1990). *The firm, the market, and the law*. Chicago: University of Chicago Press.

COLLINS, J. (2001). *Good to great*. New York: HarperCollins.

CSIKZENTMIHALYI, M. (1990). *Flow: The psychology of optimal experience*. New York: HarperCollins.

ELBOW, P. (1998). *Writing with power: Techniques for mastering the writing process* (2nd ed.). New York: Oxford University Press.

FastTrac (2002). *Technology entrepreneurship*. Kansas City, MO: Kauffman Center for Entrepreneurial Leadership.

FRANKLIN, B. (2004). *Benjamin Franklin: The autobiography and other writings*. New York: Touchstone. (Original work published 1791)

FRITZ, R. (1991). *Creating*. New York: Fawcett Columbine.

FUKUYAMA, F. (1992). *The end of history and the last man*. New York: Free Press.

GOLDBERG, D. E. (1989). *Genetic algorithms in search, optimization, and machine learning*. Reading, MA: Addison-Wesley.

GOLDBERG, D. E. (2002). *The design of innovation: Lessons from and for competent genetic algorithms*. Boston: Kluwer Academic Publishers.

GORDON, K. E. (1993a). *The deluxe transitive vampire*. New York: Pantheon Books.

GORDON, K. E. (1993b). *The new well-tempered sentence* (rev. ed.). New York: Houghton Mifflin.

GREENWOOD, E. (1957). Attributes of a profession. *Social Work, 2*, 44–55.

HANDY, C. (1995). *Gods of management: The changing work of organizations*. New York: Oxford University Press.

HANDY, C. (1998). *Beyond certainty: The changing world of organizations* (reprint ed.). Boston: Harvard Business School Press.

HARRIS, Jr., C. E., PRITCHARD, M. S., & RABINS, M. J. (1995). *Engineering ethics: Concepts and cases*. Belmont, CA: Wadsworth.

HAYEK, F. (1945). The use of knowledge in society. *American Economic Review, 35*(4), 519–530.

HAZLITT, H. (1964). *The foundations of morality*. New York: C. Van Nostrand.

HAZLITT, H. (1979). *Economics in one lesson*. New York: Arlington House.

HOGAN, B. with H. W. WIND (1957). *The modern fundamentals of golf*. New York: Barnes.

IEEE (1990). IEEE code of ethics. http://www.ieee.org/portal/pages/about/whatis/code.html

KATZENBACH, J. R., & SMITH, D. K. (2003). *The wisdom of teams: Creating the high-performance organization*. New York: HarperCollins.

KOUZES, J. M., & POSNER, B. Z. (2003). *The leadership challenge* (3rd ed.). San Francisco: Jossey-Bass.

KNUTH, D. E., LARRABEE, P., & ROBERTS, P. M. (1989). *Mathematical writing*. Washington, DC: Mathematical Association of America.

LAYTON, E. T. (1990). *The revolt of the engineers: Social responsibility and the American engineering profession*. Baltimore: Johns Hopkins University Press.

MCGREGOR, D. (1985). *The human side of enterprise* (25th anniversary ed.). New York: McGraw-Hill.

MASLOW, A. (1987). *Motivation and personality* (3rd ed.). New York: HarperCollins.

MATHES, J. C., & STEVENSON, D. W. (1991). *Designing technical reports* (2nd ed.). New York: Wiley.

MILGRAM, S. (1963). Behavioral study of obedience. *Journal of Abnormal and Social Psychology, 67*, 371–378.

MILL, J. S. (1993). *Utilitarianism*. London: Everyman. (Original work published 1861)

MILLER, G. J. (1992). *Managerial dilemmas: The political economy of hierarchy*. Cambridge, UK: Cambridge University Press.

NSPE (2003). NSPE code of ethics for engineers. http://www.nspe.org/ethics/eh1-code.asp

OSBORN, A. F. (1963). *Applied imagination*. New York: Scribners.

PARK, C. S. (2002). *Contemporary engineering economics* (3rd ed.). Upper Saddle River, NJ: Prentice Hall.

PETERSON, C., & SELIGMAN, M. E. P. (2004). *Character strengths and virtues: A handbook and classification*. New York: Oxford University Press.

PINK, D. H. (2001). *Free agent nation: How America's new independent workers are transforming the way we live*. New York: Warner Books.

PORTER, M. (1980). *Competitive strategy*. New York: Basic Books.

PORTER, M. (1985). *Competitive advantage*. New York: Free Press.

ROTTER, J. B. (1954). *Social learning and clinical psychology*. New York: Prentice-Hall.

ROTTER, J. B. (1966). Generalized expectancies for internal versus external control of reinforcement. *Psychological Monographs, 80*. (Whole No. 609)

SCHOLTES, P. R. (1998). *The leader's handbook: Making things happen, getting things done*. New York: McGraw-Hill.

SCHOLTES, P. R., JOINER, B. L., & STREIBEL, B. J. (2003). *The TEAM® handbook* (3rd ed.). Madison, WI: Oriel Incorporated.

SEARLE, J. R. (1995). *The construction of social reality*. New York: Free Press.

SELIGMAN, M. E. P. (1998). *Learned optimism*. New York: Free Press.

SELIGMAN, S. E. P. (2002). *Authentic happiness*. New York: Free Press.

SELIGMAN, M. E. P., MAIER, S. F., & GEER, J. (1968). The alleviation of learned helplessness in dogs. *Journal of Abnormal Psychology, 73*, 256–262.

SMITH, A. (1937). *An inquiry into the nature and causes of the wealth of nations*. New York: Modern Library. (Original work published 1776)

SMITH, A. (1984). *The theory of moral sentiments*. Indianapolis: Liberty Fund. (Original work published 1759)

SOWELL, T. (1996). *Knowledge and decisions*. New York: Basic Books.

SOWELL, T. (2004). *Basic economics: A citizen's guide to the economy* (rev. ed.). New York: Basic Books.

STACK, J. (1992). *The great game of business*. New York: Doubleday.

STANLEY, T. (2001). *The millionaire mind*. Kansas City, MO: Andrews Mc Meel.

STRUNK, W., Jr., & WHITE, E. B. (2000). *The elements of style* (4th ed.). New York: Macmillan.

TIMMONS, J. A. (1999). *New venture creation: Entrepreneurship for the 21st century*. Boston: Irwin McGraw-Hill.

TOBIAS, A (2005). *The only investment guide you'll ever need*. New York: Harvest Books.

TOWNSEND, R. (1984). *Further up the organization*. New York: Knopf.

TRACY, J. A. (1999). *How to read a financial report for managers, entrepreneurs, lenders, lawyers, and investors* (5th ed.). New York: Wiley.

University of Chicago Press (2003). *The Chicago manual of style* (15th ed.). Chicago: Author.

VINCENTI, W. G. (1990). *What engineers know and how they know it: Analytical studies from aeronautical history*. Baltimore: Johns Hopkins University Press.

WILSON, J. Q. (1993). *The moral sense*. New York: The Free Press.

Index